BURLEIGH DODDS SCIENCE: INSTANT INSIGHTS

NUMBER 08

Heat stress in dairy cattle

I0130557

burleigh dodds
SCIENCE PUBLISHING

Published by Burleigh Dodds Science Publishing Limited
82 High Street, Sawston, Cambridge CB22 3HJ, UK
www.bdspublishing.com

Burleigh Dodds Science Publishing, 1518 Walnut Street, Suite 900, Philadelphia, PA 19102-3406, USA

First published 2021 by Burleigh Dodds Science Publishing Limited
© Burleigh Dodds Science Publishing, 2021. All rights reserved.

This book contains information obtained from authentic and highly regarded sources. Reprinted material is quoted with permission and sources are indicated. Reasonable efforts have been made to publish reliable data and information but the authors and the publisher cannot assume responsibility for the validity of all materials. Neither the authors nor the publisher, nor anyone else associated with this publication shall be liable for any loss, damage or liability directly or indirectly caused or alleged to be caused by this book.

No part of this publication may be reproduced, stored in a retrieval system or transmitted in any form or by any means electronic, mechanical, photocopying, recording or otherwise without the prior written permission of the publisher.

The consent of Burleigh Dodds Science Publishing Limited does not extend to copying for general distribution, for promotion, for creating new works, or for resale. Specific permission must be obtained in writing from Burleigh Dodds Science Publishing Limited for such copying.

Permissions may be sought directly from Burleigh Dodds Science Publishing at the above address. Alternatively, please email: info@bdspublishing.com or telephone (+44) (0) 1223 839365.

Trademark notice: Product or corporate names may be trademarks or registered trademarks and are used only for identification and explanation, without intent to infringe.

Notice
No responsibility is assumed by the publisher for any injury and/or damage to persons or property as a matter of product liability, negligence or otherwise, or from any use or operation of any methods, products, instructions or ideas contained in the material herein.

British Library Cataloguing in Publication Data
A catalogue record for this book is available from the British Library

ISBN 978-1-78676-933-6 (Print)
ISBN 978-1-78676-934-3 (ePub)

DOI 10.19103.9781786769343

Typeset by Deanta Global Publishing Services, Dublin, Ireland

Contents

© Burleigh Dodds Science Publishing Limited, 2021. All rights reserved.

© Burleigh Dodds Science Publishing Limited, 2021. All rights reserved.

Chapter 1

Advances in dairy cattle breeding to improve heat tolerance

Thuy T. T. Nguyen, Agriculture Victoria, Australia

1 Introduction

Global warming is posing a potential challenge for all food production sectors, and the dairy industry is no exception in this regard. During lactation, dairy cows generate a high amount of internal metabolic heat and as a consequence become particularly sensitive to hot environments (Kadzere et al., 2002). When the environmental temperature and humidity exceeds the comfort level (i.e. thermoneutral zone), cows have difficulties in dissipating the metabolic heat and become prone to heat stress (Kadzere et al., 2002; Berman, 2011). While heat stress is a common problem encountered in the tropics, it becomes an increasingly important issue for dairy cattle in temperate climates because of climate change (Renaudeau et al., 2012). The Intergovernmental Panel on Climate Change (IPCC) projected that by the end of the twenty-first century the global mean surface temperature is likely to exceed by 1.5°C relative to pre-industrial level (IPCC, 2014). The warming will continue beyond 2100 unless significant efforts to reduce the greenhouse gas emissions are put in place. Heatwaves will become more frequent and be of longer duration. These projections indicate that the impacts of climate change to the dairy industry will become increasingly significant in the future. As such, development of strategies to mitigate the impacts of heat stress becomes an important and immediate task for the dairy industry in many countries.

http://dx.doi.org/10.19103/AS.2019.0058.16
© Burleigh Dodds Science Publishing Limited, 2020. All rights reserved.

The hot and humid weather conditions have significant direct impacts on the health and productivity of dairy cattle. When environmental parameters such as temperature and humidity exceed comfort levels (e.g. temperature of 22°C and relative humidity of 50%), dairy cows reduce their feed intake and milk production (Silanikove, 1992; St-Pierre et al., 2003; Garner et al., 2017). In some instances, health, fertility, immune function and animal welfare could also be compromised (Hansen and Aréchiga, 1999; Kadzere et al., 2002; Dikmen and Hansen, 2009; Tao and Dahl, 2013).

Sensitivity of dairy cows to heat stress is influenced by many factors. For example, some breeds are more tolerant to heat stress than others (Hansen, 2004; Davis et al., 2017), and multiparous cows are more resistant to heat stress compared to the primiparous counterparts (Gantner et al., 2017). Lactating dairy cows are more sensitive to hot weather conditions compared to non-lactating cows because the former generates more metabolic heat for milk production (West, 2003). Equally, animals in early lactation are more sensitive to heat stress compared to late-lactating cows (Kadzere et al., 2002). High milk-producing cows are more susceptible to heat stress, due to the increase in the metabolic heat output (Purwanto et al., 1990). The selection for high-producing cows, therefore, also resulted in the selection of animals which are more vulnerable to heat stress (Nguyen et al., 2017a).

The economic loss brought about by heat stress can be substantial. It is estimated that the annual economic loss to the US dairy industry attributed to heat stress, as a result of decreased milk production, compromised reproduction and increased culling, is approximately £550 million (St-Pierre et al., 2003). On the loss of production alone, cows in Australia are estimated to lose up to 12% of their annual production, valued at approximately £170 million (DairyBio, 2018). In the United Kingdom, it is projected that, particularly in the South West England region, the economic loss due to direct impacts of heat stress in the dairy industry may reach £13.4 million in an average year and £33.8 million in extreme years (Fodor et al., 2018).

Efforts to mitigate the impacts of heat stress could be multifaceted, involving management, nutrition and genetics. Provision of shades, fans and sprinklers on hot days or changing milking time to a cooler period of the day can significantly reduce the level of stress on animals. It is also generally recommended that during periods of intense heat a carefully balanced diet with less fibre and more concentrates, such as slowly fermentable grains, can be used to reduce dietary heat load and to ensure energy intensity for milk production (Vermunt and Tranter, 2011; Conte et al., 2018; Gonzalez-Rivas et al., 2018). While environmental and nutritional manipulation can be effective, the use of genetically improved heat-tolerant animals can provide a complementary and a longer-term solution to the problem. Recent advances in this field provide new tools for farmers to utilise genetic resources that are known to adapt well to hot conditions (Davis et al., 2017) or to select animal with better heat tolerance if they are farming in areas with hot and humid conditions or under the threat of global warming (Garner et al., 2016; Nguyen et al., 2016, 2017a).

This chapter aims to describe current advances in dairy cattle breeding to improve heat tolerance. These technologies range from traditional cross-breeding methods to generate a hybrid that carries characteristics of both tropically adapted animals and high-producing temperate breeds of dairy cattle, to the use of modern genomic tools to expedite the selection of highly tolerant animals for the changing climate. Future directions on the improvement of data, models and the use of multiomics technologies in the improvement of heat tolerance are also discussed.

© Burleigh Dodds Science Publishing Limited, 2020. All rights reserved.

2 Utilisation of tropically adapted breeds for cross-breeding

The majority of milk produced today comes from dairy breeds of two subspecies: the zebu cattle (or humped cattle, *Bos taurus indicus*) and the taurine cattle (or humpless cattle, *B. taurus taurus*). These two subspecies were domesticated independently in India and in the Near East, respectively (Bradley et al., 1996; Beja-Pereira et al., 2006). The division between the zebu and the taurine lineages is estimated to have occurred from a common ancestor between several hundred thousand (Bradley et al., 1996; MacHugh et al., 1997) and 2 million (Hiendleder et al., 2008) years ago. Because of their long separation and different selection pressures, the zebu and taurine cattle differ substantially in many traits such as heat tolerance and resistance to parasites (Utech and Wharton, 1982; Madalena et al., 1990; Wambura et al., 1998; Hansen, 2004; Glass et al., 2005), both of which are key attributes for survival in a tropical environment.

The superiority in thermotolerance in zebu cattle is evident at both physiological and cellular levels, and is attributed to a lower heat production as well as an increased ability for heat dissipation (Hansen, 2004). Zebu cattle breeds generally have lower milk yield compared to taurine breeds, and therefore the amount of metabolic heat generated is relatively less (West, 2003). In a study to compare heat tolerance between breeds of taurine (Holsteins and Jerseys) and the zebu × taurine cross-breeds (Red Sindhi × Holstein), Johnston et al. (1958) found that the total heat production and heat production per unit of surface area in the cross-breeds were lower than those in the pure Holsteins and Jerseys. In addition, zebu cattle have a better ability to dissipate heat as they have higher density and larger sweat glands (Dowling, 1955; Nay and Heyman, 1956; Pan, 1963), accompanied with a better blood flow to the periphery, thereby enhancing heat loss via the skin surface (Finch, 1985). While the thick hair coat in the taurine cattle is an important adaptation to a cold climate, it is disadvantageous in hot conditions as it restricts airflow and evaporative loss (Seif et al., 1979; Berman, 2005). The zebu cattle, conversely, have a shorter hair coat and hence facilitate heat loss.

Due to the high thermotolerance and parasite resistance properties, zebu breeds are often used for tropical dairying. These breeds, however, generally have lower milk yield and later first calving dates compared to taurine breeds (Davis et al., 2017). Therefore, cross-breeding between the local breeds with high producing taurine such as Holsteins or Jerseys are often practised in tropical countries to create a hybrid that maintains the thermotolerance characteristics of zebu and at the same time improving milk yield via heterosis (Galukande et al., 2013).

Apart from zebu cattle, hot weather adapted breeds are also found among the taurine, for example the Senepol cattle. It has been reported that Senepol cattle have a similar level of thermotolerance to the zebu (Brahman) cattle (Hammond and Olson, 1994; Hammond et al., 1996). The F1 cross-breeds between Senepol and temperate breeds are also as thermotolerant as Brahman and Brahman cross-breeds (Hammond and Olson, 1994; Hammond et al., 1996, 1998). The superiority in heat tolerance in Senepol cattle is linked to its 'slick hair' coat that is determined by a dominant gene, the 'slick hair' gene (Olson et al., 2003). In fact, the casual mutation for the 'slick' hair is identified as a single base deletion in the prolactin receptor region of chromosome 23, causing a frameshift that truncates 120 amino acids from the long isoform of the receptor (Littlejohn et al., 2014).

© Burleigh Dodds Science Publishing Limited, 2020. All rights reserved.

Introgression of the 'slick hair' mutation into dairy breeds such as Holsteins shows promise (Dikmen et al., 2014; Davis et al., 2017). In the United States, Holsteins introgressed with 'slick hair' have superior thermotolerance in comparison to 'normal' Holsteins, showing lower vaginal temperature and less decline in milk yield in summer, as well as a lower rectal temperature and respiration rate, and higher sweating rate during heat stress (Dikmen et al., 2014). In New Zealand, there is an ongoing 5-year program to introgress the 'slick' variant from Senepol cattle to the New Zealand dairy breeds, aiming to produce homozygous 'slick' bulls that would have at least 75% New Zealand dairy genetics (Davis et al., 2017). If these 'slick' bulls are very high in breeding worth (New Zealand national measure of genetic merit) then it is a very effective way to propagate heat tolerance in a single step. This is a slow process but offers a potential adoption from the dairy industry in the tropics.

The finding of the single base mutation with a major effect on thermotolerance also opens an opportunity for genome editing to improve this important trait. Although genome editing to date is still limited in cattle, it has proven successful in several cases such as the introduction of myostatin for double muscling in meat production (Luo et al., 2014; Proudfoot et al., 2015), production of polled (hornless) cattle (Tan et al., 2013; Carlson et al., 2016), elimination of allergen in milk (Yu et al., 2011) or producing disease-resistant animals (Liu et al., 2013, 2014; Wu et al., 2015). From a tropical dairy perspective, it is envisaged that the genome of high producing breeds such as Holsteins can be edited to include tropically adapted variants such as 'slick' and those associated with coat colour and sweating rate, making the animals more adaptable to hot and humid environments (Davis et al., 2017). The regulatory policies for producing such animals, however, vary among countries, and the acceptance of such products ultimately depends on the wider public.

3 Selection for heat-tolerant dairy cattle

3.1 Heat tolerance phenotypes

One way to improve the ability to cope with hot and humid environments in dairy cattle is to continuously select for the trait. This is different to the single gene modification approach described above where heat tolerance is selected based on the additive effects of multiple genes across the genome. It is, however, a direct phenotype for heat tolerance, such as rectal temperature, that is either currently unavailable or difficult to collect on a large-scale routine evaluation. Ravagnolo and Misztal (2000) proposed the use of the rate of production decline with increasing heat stress as a proxy to heat tolerance. Production in heat-tolerant cows drops more slowly in response to increasing heat stress compared to cows that are more susceptible to heat stress. Figure 1 provides an example of how cows perform differently under heat stress conditions. Under thermoneutral conditions, that is within the comfort zone, both cow A and cow B produce a similar amount of milk. However, when environmental heat load increases beyond the thermoneutral zone (the upper critical zone), that is heat stress conditions, cow B produces less milk than cow A. This indicates that cow A is more tolerant to heat stress compared to cow B.

The use of the rate of production decline during heat stress as an intermediate phenotype has an advantage that no additional data has to be collected, and the large production

© Burleigh Dodds Science Publishing Limited, 2020. All rights reserved.

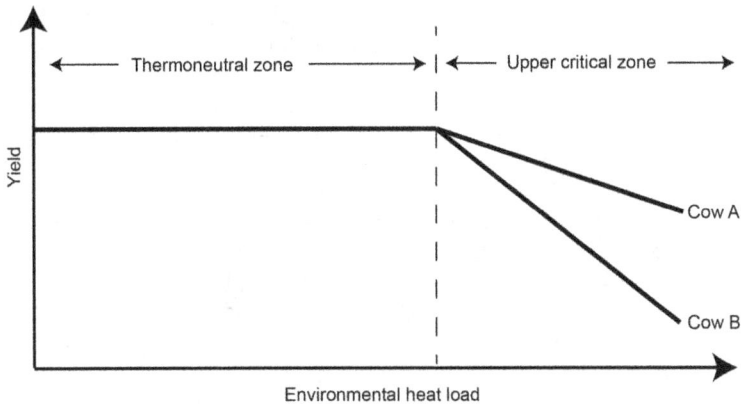

Figure 1 The rate of decline in milk yield with increasing heat stress as a proxy for heat tolerance. Source: adapted from Ravagnolo and Misztal (2000).

dataset used for national genetic evaluations can be employed to study heat tolerance. This approach has been widely applied in a number of recent studies (Hayes et al., 2003; Bohmanova et al., 2005; Finocchiaro et al., 2005; Brügemann et al., 2011; Hammami et al., 2013; Bernabucci et al., 2014; Carabaño et al., 2014; Nguyen et al., 2016). Heat tolerance, in this case, the rate of production decay as environmental heat load increase, possesses a significant level of genetic variation and is genetically in antagonism with milk production (Bohmanova et al., 2005; Nguyen et al., 2016). Also, genotype by environment exists between production under thermoneutral conditions and that in heat-stress environments (Hayes et al., 2003; Misztal et al., 2006). That means a bull which has some daughters producing high milk yields in cool climates and other daughters poorly performing in heat stress conditions is less tolerant than a bull whose daughters produce well in all environments. Although performance decay under heat stress is not an ideal phenotype for heat tolerance, it is the most cost-effective measure for the trait available to date. New tools are available to improve the phenotypes for heat tolerance and are discussed in Section 4.2.

3.2 Environmental heat load

Environmental heat load is a product of many factors such as temperature, humidity, wind speed and solar radiation. Thermoregulation in cattle is affected largely by air temperature and relative humidity. The temperature–humidity index (THI) combines temperature and humidity into a single parameter and is the most commonly used indicator of environmental heat load. Weather records from public meteorological stations are often used for this purpose.

There are many formulae to compute THI and the use of these formulae varies among studies. Bohmanova et al. (2007) tested seven different indices and found that those with more weight on humidity are a better indicator in humid climates, whereas those with heavier weighting on temperature are more appropriate for semiarid regions. These authors also found that the threshold, the point at which the production begins

© Burleigh Dodds Science Publishing Limited, 2020. All rights reserved.

to decline, is different between indices. The time point at which the temperature and humidity values are used to calculate THI also varies. For example, some studies use the daily maximum temperature and minimum humidity in the calculations (Ravagnolo and Misztal, 2000; Bernabucci et al., 2014), whereas others use mean daily temperature and mean humidity (Hammami et al., 2015), or average hourly THI (Aguilar et al., 2010b; Nguyen et al., 2016). Even when the same manner of expression of temperature (maximum or average) is used, some keep to the Celsius scale (Ravagnolo and Misztal, 2000), while others transform to the Fahrenheit scale before calculating THI (Bernabucci et al., 2014). Moreover, the time lag between the heat event and the decline of production are also different between regions (Aguilar et al., 2010b; Nguyen et al., 2016). All these variations highlight the fact that the threshold value depends on the THI formula used, as well as what daily values of temperature and humidity, unit of temperature and the time lag between the heat event and the milk recording day. Moreover, the threshold can change over time as the emphasis for selection on production may lead to a decline in heat tolerance (Zimbelman et al., 2009; Nguyen et al., 2017a), that is a lower threshold value is expected. This threshold value is important in modelling for heat tolerance and as such an estimation of the value is needed for every study.

3.3 Modelling and estimation of genetic parameters for heat tolerance

In animal breeding, a reaction norm, or a mathematical function that relates the mean phenotypic expression of a genotype to an environmental gradient, is widely used to study the responses of animals to their environment (de Jong and Bijma, 2002; Hayes et al., 2003; Calus et al., 2005). In a reaction norm model, environmental sensitivity is defined as the first derivative of the reaction norm function (Kolmodin et al., 2002; Hayes et al., 2016). That means in the case of heat tolerance, the slope of the linear reaction norm is the measure of sensitivity of an individual towards an increase in THI (Hayes et al., 2003, 2009b; Haile-Mariam et al., 2013; Hammami et al., 2015).

In linear reaction norm models, a heat load function needs to be created. This function shows that production is unaffected when a cow is in the thermoneutral zone, and starts to decline when THI exceeds the threshold. The heat load function can be written as the following:

$$f(THI) = \begin{cases} 0 & \text{if } THI \le \text{threshold} \\ (THI\text{-threshold}) & \text{if } THI > \text{threshold} \end{cases}$$

where THI is the THI value associated with the herd test record, and the threshold is the THI at which production begins to decline (Ravagnolo and Misztal, 2000).

Although a reaction norm model can be used in combination with a repeatability model, it is recommended that a random regression implementation of reaction norm is better in capturing the dynamics of heat stress throughout lactation (Ravagnolo and Misztal, 2000). A general formulation of a random regression model for the case of a linear reaction norm is:

$$y_{ij} = \mu + \delta_i + (\beta + \Delta_i) X_{ij} + e_{ij}$$

© Burleigh Dodds Science Publishing Limited, 2020. All rights reserved.

where y_{ij} is the value of dependent variable (e.g. milk yield) and X_{ij} is the value of the environment (e.g. a linear reaction norm of f(THI)) for measurement i from individual j; μ and β are the means of intercept and slope, respectively; δ_i and Δ_i are the individual variations from these means, respectively; δ_i and Δ_i have a mean of zeros and a (co)variance structure defined by a matrix **C** which is to be estimated. **C** is a 2 × 2 matrix with σ_δ^2 (the variance of intercept) and σ_Δ^2 (the variance of the slope) on the diagonal, and $\sigma_{\delta,\Delta}$ (the covariance between intercept and slope) as the off-diagonal element; and e_{ij} is the error term, $e_{ij} \sim N(0, \sigma_{e_{ij}}^2)$. Assuming S_i is a n × 1 vector of breeding values for intercept or slope ($i = 0$ is the intercept and $i = 1$ is the slope). The variance of S is $\mathrm{var}(S) = \mathbf{A} \otimes \mathbf{C}$, where **A** is the relationship matrix derived from the pedigree.

In linear reaction norm models, it is assumed that there is a common threshold value for all individuals. In fact, threshold values vary among individuals. For example, heat-tolerant animals start declining production at a higher temperature and humidity than heat-susceptible counterparts and vice versa. In order to take threshold estimation into account, Sánchez et al. (2009) proposed a method to simultaneously estimate a threshold and a slope for each animal. These authors found that both threshold and slope have large genetic components and the genetic correlation between them is 0.95, indicating that an animal with a higher threshold of heat stress would also have lower rate of production decline and vice versa. As such, selection of one component of heat tolerance would lead to a favourable response for the other. Sánchez et al. (2009) also pointed out that the process of estimating both threshold and slope could experience slow mixing and convergence rates and suggested that a simple model assuming a common threshold at which THI affects milk production across animals. Although this approach may result in a bias in estimated slopes, it could still provide reasonable estimates of breeding value that capture most of the genetic variation for heat tolerance.

Within the range of THI normally encountered in dairy environments, it is reasonable to assume that reaction norms are linear. However, it is also possible to consider nonlinear reaction norm models using polynomial functions, which may show a better fit of the phenotype and genetic variation of production traits along the range of THI (Brügemann et al., 2011; Carabaño et al., 2014). Higher degree of polynomials of random regression gives a smoother pattern of changes in the traits affected by heat stress (e.g. milk yield); and do not require a threshold value as in the case of linear reaction norms (Carabaño et al., 2014, 2016). However, when higher order terms of polynomials such as quadratic and cubic are used, associated variance components of coefficients can be difficult to estimate (Hayes et al., 2016).

Selection of heat-tolerant animals based on the slope obtained from reaction norm models are likely to compromise production due to the unfavourable genetic relationships between the two traits. Carabaño et al. (2017) suggested that the eigen decomposition of the co(variance) matrix among random regression coefficients can be used to decompose the variability in the shapes of response to heat load in independent components. For example, Carabaño et al. (2014) used a cubic polynomial reaction norm and found that the first principal component was associated with the overall level of production, whereas the remaining three principal components were independent from the first and associated with changes with increasing heat load. This meant that the latter three components could be used to select for heat-tolerant animals without compromising production. Although this approach is worth exploring, the authors cautioned that further checking

© Burleigh Dodds Science Publishing Limited, 2020. All rights reserved.

on the consistency of decomposition is warranted and that the biological interpretation of principal components is difficult.

3.4 Genomic selection

With the dawn of the current millennium, genomic selection has revolutionised the way in which animal and plant breeding is conducted (Meuwissen et al., 2001, 2016). Traditional selection methods based on pedigree information can now be replaced by or used in combination with genomic relationship between individuals (Aguilar et al., 2010a). Given that sequencing and genotyping costs are declining, and most of national dairy genetic evaluation programmes would have an existing large population of genotyped cows and bulls, the estimation of genomic breeding values (GEBV) for newly emerging traits such as heat tolerance evaluated using freely available milk production measurement would incur very minimal costs. With traditional breeding, a bull would have to have his progeny tested, that is his daughters start to have records on milk production, before a selection decision on him can be made. With genomic selection, bulls and heifers can be selected as young as at birth, provided that a genotype can be obtained. In practice, dairy bulls have already been, and are, increasingly selected for a range of traits based on their GEBV rather than progeny testing. Genomic selection for heat tolerance therefore can also be implemented. With an increasing demand in importing heifers from temperate dairying countries to tropical regions, a GEBV for heat tolerance for these heifers becomes valuable. It is noteworthy that there could be variation in production among sire's daughters at different heat stress levels and such genotype by environment interaction needs to be taken into account when selection decision is made.

To date, GEBV for heat tolerance is only implemented in Australia (by DataGene, Bundoora, Australia) and available for all genotyped Holstein and Jersey cattle (Nguyen et al., 2016, 2017a). The development and related process to implement this genomic breeding value are described in Section 3.5.

3.5 Case study: the development, validation and implementation of heat tolerance GEBV in Australia

Combining weather data from publicly available weather stations and herd test records over a 14-year period, Nguyen et al. (2017a) then used a linear reaction norm model to estimate the rate of decline in milk, fat and protein yield per unit increase of THI above the threshold where heat stress affects production. Hourly THI was calculated using the method of Yousef (1985), then averaged for the test day and 4 days before the test day to represent the environmental heat load of that test day. The threshold value was set to THI = 60 as determined by Hayes et al. (2003). These estimated rates of decline were used as pseudo-phenotypes for heat tolerance in Holsteins and Jerseys. For each breed, a prediction equation was derived using the genomic best linear unbiased prediction method. The reference population included 2 236 sires and 11 853 cows in Holsteins, and 506 sires and 4 268 cows in Jerseys. Each animal in the reference population had an estimated slope and a medium-density (~50 K SNP) genotype. Heat tolerance GEBV was then determined for other animals with genotypes. The reliability of heat tolerance GEBV was 38% in both breeds.

© Burleigh Dodds Science Publishing Limited, 2020. All rights reserved.

Table 1 Correlations between direct genomic values (DGV) of heat tolerance and Australian breeding values (ABV) of other key traits of Holstein bulls

DGV for heat tolerance	ABV					
	Milk yield	Fat yield	Protein yield	Fertility[a]	Residual survival[a]	Feed saved[a]
Milk	−0.65	−0.20	−0.50	0.11	−0.06	0.17
Fat	−0.12	−0.75	−0.31	0.10	0.06	0.17
Protein	−0.44	−0.13	−0.65	0.13	−0.02	0.15

[a] The fertility ABV is expressed as 6-week in-calf rate and derived from a five-trait model that includes calving interval, lactation length, first service non-return rate, days to first service and pregnancy.
The residual survival ABV is survival ABV adjusted for traits known to influence a cow's herd life.
Feed saved is the number of kilograms of feed that are saved from improvements in residual feed intake and maintenance derived from body weight.

To further validate the heat tolerance GEBV, Garner et al. (2016) predicted heat tolerance GEBV for 390 Holstein heifers, then selected 24 extreme predicted heat-tolerant and 24 extreme predicted heat-susceptible heifers for a 4-day heat challenge in climate-controlled chambers. The trials were conducted at agriculture research facilities in Ellinbank, Victoria, Australia. The predicted heat-tolerant group showed significantly less decline in milk production, lower rectal and intravaginal temperatures than the predicted heat-susceptible group. This indicates that the heat tolerance GEBV enables selection for cattle with better tolerance to heat stress.

Heat tolerance GEBV was expressed by applying economic weightings for decline in milk, fat and protein, as used by the industry. Within each breed, the heat tolerance GEBV was standardised to have a mean of 100 and standard deviation of 5. Table 1 presents the correlation between heat tolerance GEBV and some key traits in Australian Holsteins. Heat tolerance GEBV was found to be unfavourably correlated to production and favourably correlated to fertility and feed saved. The HT GEBV was released by DataGene in December 2017.

4 Future perspectives on breeding for heat tolerance

Implementation of heat tolerance selection is in its early stages and the outcome may not be seen for several generations. Significant research results have been achieved to date in terms of exploring different types of phenotypes for heat tolerance and model development. Strategies such as utilisation of publicly available weather information and indirect measure of phenotypes based on existing production data appear cost-effective but could be improved. The new 'omics' era opens opportunities to explore for a more precise measure of heat tolerance than production decay, and in combination with genomic selection the rate of genetic gain is expected to increase. Reliability of genomic prediction is also expected to improve when more biological information on heat tolerance can be incorporated to derive prediction equations, and more genotyped cows are included in the reference population. The following sections aim to discuss these aspects in the context of future breeding for heat tolerance.

© Burleigh Dodds Science Publishing Limited, 2020. All rights reserved.

4.1 Climate data

Publicly available historical weather records have been widely used to calculate environmental heat load. These weather records can be obtained from meteorological agencies at a minimal cost. Such data from the nearest station to a herd can be used to represent the herd climate. Freitas et al. (2006) compared the use of on-farm recorded temperature and humidity with those recorded at available weather stations (3 to >400 km distance from the farm). The study found that weather station data can be as informative as on-farm measurements (Misztal et al., 2006). However, in countries or regions where weather stations are scarce and/or with uneven terrains, an alternative is to have a mini-weather station installed on farm. Information captured from these on-farm stations are extremely valuable, particularly for herds with good phenotype records such as the Genomic Information Nucleus (Ginfo) herds in Australia (Pryce et al., 2017). Cows from these Ginfo herds have frequent herd test records and are being genotyped to be included in the reference population for genomic prediction. Some meteorological agencies are able to conduct high-resolution reanalyses for the atmosphere over a number of years or decades. For example, in Australia, the Bureau of Meteorology Atmospheric high-resolution Regional Reanalysis for Australia is downscaling historical hourly temperature and humidity to a 12-km resolution, and in some regions it is even at as a fine scale as 1.5-km resolution. Access to those downscaled climate data is expected to result in a more accurate climate presentation for the herds compared to the weather stations data, and allow cows with good herd testing records but are distant from weather stations contributing to heat tolerance analysis. It is envisaged that with better climate representation of the environment, more variance on heat tolerance could be captured.

4.2 Potential new predictors for heat tolerance phenotypes

The use of herd test records for production traits to estimate heat tolerance has an advantage that no additional data are required, and can be seen as a cost-effective way to achieve a breeding value for heat tolerance. This approach, however, has some certain limitations. Herd test data could capture only a fraction of heat stress compared to more frequent measurements (Misztal et al., 2006). Herd testing provides only a few records per cow per year and therefore accounting for past events that influence the test day milk yield is difficult. In this regard, obtaining production data from automatic milking systems become a plausible alternative as these data can be recorded daily. These systems are often available for large herds and therefore production data can be obtained for a large number of cows. It is unfortunate that records from such herds might have not yet been entered into the national herd test systems in many countries and currently excluded from analyses. Much work is required to standardise information from these systems before it can be used effectively.

Identifying a more precise measure of heat tolerance than the production decay on hot days is an appropriate direction. New technologies such as metabolomics, transcriptomics, genomics and phenomics provide an excellent opportunity in this regard. Despite this, core body temperature is considered the most appropriate measure of heat tolerance and is among the most important phenotypes in livestock animals as it is related to productivity, health and reproduction (St-Pierre et al., 2003). However, collection of such data currently cannot be achieved on a daily basis and on a large scale. Collection of data

© Burleigh Dodds Science Publishing Limited, 2020. All rights reserved.

on other indicators of heat stress such as respiration rate (Collier et al., 2006), sweating rate (Dikmen et al., 2014) and blood flow measurement (Honig et al., 2016) are also labour intensive. New development in automated body temperature monitoring technologies allows body temperature data to be measured in a real-time manner (Koltes et al., 2018). Technologies such as temperature-sensing ear tags, rumen-reticular boluses, intravaginal and intrarectal thermal-sensor devices and thermal imaging are commonly used. But these are more appropriate in research environments rather for commercial herds due to the costs involved. Even these data are collected for a small group of cows; it is a valuable trait and can be used in conjunction with other indicators to improve estimation of heat tolerance. Note that the use of sensor tags is growing at a fast pace in dairy cattle (mainly for oestrus detection), but this could be an example of another trait derived as a by-product of mainstream management tools. Most experiments on heat stress to date require climate-controlled chambers to induce heat stress. Al-Qaisi et al. (2019) proposes the use of electric heat blanket as an alternative and cost-effective method to measure heat stress in dairy cows.

Metabolomic tools such as integrated ^{1}H nuclear magnetic resonance and ultra-fast liquid chromatography-mass spectrometry are powerful in identifying low-molecular weight metabolites associated with physiological changes. Metabolites derived from plasma and milk samples using these techniques were able to distinguish heat-stressed and normal lactating dairy cows, and as such can be potential biomarkers of heat stress (Tian et al., 2015, 2016). Examples of these markers include lactate, pyruvate, creatine, acetone, β-hydroxybutyrate, trimethylamine, oleic acid, linoleic acid, lysophosphatidylcholine 16:0 and phosphatidylcholine 42:2 (Tian et al., 2015, 2016). Because of a high correlation between milk and plasma in the levels of these metabolites, analysing these biomarkers from milk samples would be sufficient as milk samples are relatively more convenient to be collected (Tian et al., 2016).

Proteomics offers a powerful approach to discover proteins and pathways that are crucial for stress responsiveness and tolerance. Min et al. (2016) used the isobaric tags for relative and absolute quantification (iTRAQ) approach to analyse plasma proteomes of heat-stressed and normal cows, and identified 85 proteins that were differentially abundant in the two groups. Examples of such proteins are adiponectin and mannose-binding protein C which were upregulated, and selenoprotein P which were downregulated in heat-stressed cows. Proteomic studies in relation to heat stress in dairy cows are still in its infancy but in combination with other technologies could potentially be used to identify markers associated with heat stress that could aid management of heat stress and also serve as intermediate phenotypes for genetic selection (Min et al., 2017).

One growing research area with great potential practical application is the use of predicted phenotypes using mid-infrared (MIR) spectroscopy. MIR is routinely used to determine milk composition; and now expanding its application to predict a range of functional, novel and difficult or expensive-to-measure traits (Egger-Danner et al., 2015). Examples of research on using MIR and blood biomarkers to predict such traits include methane and methane density (Kandel et al., 2013; Vanlierde et al., 2016), fertility (Bastin et al., 2012; Gengler et al., 2013) and metabolic status (Roberts et al., 2012; van der Drift et al., 2012). MIR predicted heat tolerance is recently examined and has shown to be of considerable promise (Hammami et al., 2015), and this has a potential to provide cost-effective alternative phenotypes for heat tolerance.

© Burleigh Dodds Science Publishing Limited, 2020. All rights reserved.

4.3 Improving the reliability of genomic prediction

Currently genomic prediction for heat tolerance using medium-density SNP genotypes (~50 K SNP) can be achieved with a reliability of 38% in pure Holsteins and Jerseys (Nguyen et al., 2017b). In order to increase the rate of genetic gain, this reliability can be improved by several ways. Genomic prediction requires a large reference population to achieve a reasonable reliability. This reference population traditionally only includes males. Increasingly females with better phenotypic records are genotyped and included in the reference set. As of April 2016, 87% of the Australian reference population are female (17 108 Holsteins and 3 347 Jerseys), contributing to an increase in reliability of GEBV by 5.8% and 2.5% for young genotyped Holstein and Jersey bulls, respectively (Pryce et al., 2018). This female reference population is expanding to reach approximately 60 000 cows and it is envisaged that the reliability of GEBV will continue to increase for many traits, including heat tolerance (Nguyen et al., 2017a).

Routine genomic evaluation for heat tolerance and other traits are currently performed with multistep methods, which use daughter-trait-deviation or deregressed estimated breeding value to estimate SNP effects and then direct genomic value for selection candidates based on their genotypes (e.g. Nguyen et al., 2017a). Misztal et al. (2009) proposed that GEBV can be obtained with a single-step BLUP procedure (ssBLUP) whereby all information on phenotype, pedigree and genomic relationship matrix are used simultaneously. In this approach, the pedigree-based relationship matrix is replaced by another matrix, commonly denoted as the **H** matrix, which combines the pedigree-based relationship and the genomic relationship matrices (Aguilar et al., 2010a). Many studies using this approach reported an increase in reliability of genomic prediction compared to the traditional BLUP prediction (Aguilar et al., 2011; Li et al., 2014). This method is, however, computationally intensive due to the increased number of genotyped animals and hence the inversion of the genomic-based relationship matrix is exhausted. To overcome this, Misztal et al. (2014) suggested an approximation method, in which the genotyped individuals are divided into groups of core and noncore animals, and the regular inverse of the genomic-based relationship matrix computed only for the core animals. If the number of core animals is sufficiently large, this approach can result in a close approximation of GEBV (Masuda et al., 2016; Strandén et al., 2017). Genomic prediction for heat tolerance therefore could be potentially benefitted from this direction.

Genome-wide association studies (GWAS) can be used to identify regions of the genome that have a specific effect on heat tolerance. Previous studies on heat tolerance GWAS using medium-density (50 K SNP) genotypes identified a limited number of genes associated with the decline of production under heat stress (Hayes et al., 2009a; Macciotta et al., 2017). It is envisaged that QTL mapping using whole genome sequences and the BayesRC approach would result in a more precise detection of underlying variants and improved accuracy of genomic prediction provided that good prior biological information is available (Goddard et al., 2016; MacLeod et al., 2016). Gene expression data, such as RNA sequence data, can be valuable in this regard. To date, only a few gene expression studies have been conducted to investigate the genes involved in heat stress response in dairy cattle, and these studies report transcription profiles for only a limited number of genes (Sonna et al., 2002; Collier et al., 2008). Little information is available regarding heat-responsive genes at a global transcriptome level. Quantification of RNA sequences will allow the identification of genes and variants involved in heat stress response,

© Burleigh Dodds Science Publishing Limited, 2020. All rights reserved.

providing biological information needed for precise QTL mapping and improved reliability of genomic prediction.

5 Conclusion

The impacts of global warming on the dairy industry are an increasing concern to the industry. Devising tools and methods to mitigate such impacts is crucial for the industry worldwide. While management and nutrition help to alleviate short-term impacts, breeding for heat-tolerant dairy cows is complimentary and provides a long-term solution to the problem. To date, the most commonly used method to measure heat tolerance is the rate of decay of milk production as environmental heat load increases. In the near future, a more precise measure can be identified via the modern 'omics' technologies. It is envisaged that a faster genetic gain for heat tolerance can be achieved via identification of a better measure of heat tolerance, improved information and insights into the complexity of the mechanisms underlying the heat stress response and models used to estimate the breeding value for the trait. It is expected that the environment will continue to change, animals are expected to face more stressors than just elevated temperature, and therefore building dairy cattle which are productive and resilient to these changes is the ultimate task for the industry.

6 Acknowledgements

This work is funded by DairyBio, a joint programme between the Government of Victoria (Melbourne, Victoria) and Dairy Australia (Melbourne, Australia). The author also would like to thank DataGene (Melbourne, Australia) for providing the latest breeding values for bulls presented in Table 1. Dr. Jennie Pryce provides useful comments which greatly improve the earlier version of the manuscript.

7 Where to look for further information

7.1 Further reading

- *Environmental Physiology of Livestock* (Editors: R. J. Collier and J. L. Collier)
- *Breeding Focus 2018: Reducing Heat Stress* (Editors: S. Hermesch and S. Dominik)

7.2 Key journals

- *Journal of Dairy Science*
- *Journal of Animal Production*

7.3 Key conferences

- PAG (Plant and Animal Genome) is well attended by the members of the dairy community

© Burleigh Dodds Science Publishing Limited, 2020. All rights reserved.

- AAABG (the Association for the Advancement in Animal Breeding and Genetics), organised biannually with many sessions on dairy breeding and genetics
- EAAP (the European Federation of Animal Science), organised annually with many sessions on dairy genetics and genomics
- WCGALP (the World Congress on Genetics Applied to Livestock Production), organised every 4 year, many sessions on dairy genetics and genomics

7.4 Major international research projects

- DairyBio – http://dairybio.com.au/
- CoolCows – http://coolcows.dairyaustralia.com.au/
- GenTORE – https://www.gentore.eu/

8 References

Aguilar, I., Misztal, I., Johnson, D. L., Legarra, A., Tsuruta, S. and Lawlor, T. J. 2010a. Hot topic: a unified approach to utilize phenotypic, full pedigree, and genomic information for genetic evaluation of Holstein final score. *J. Dairy Sci.* 93(2), 743–52. doi:10.3168/jds.2009-2730.
Aguilar, I., Tsuruta, S. and Misztal, I. 2010b. Computing options for multiple-trait test-day random regression models while accounting for heat tolerance. *J. Anim. Breed. Genet.* 127(3), 235–41. doi:10.1111/j.1439-0388.2009.00842.x.
Aguilar, I., Misztal, I., Tsuruta, S., Wiggans, G. R. and Lawlor, T. J. 2011. Multiple trait genomic evaluation of conception rate in Holsteins. *J. Dairy Sci.* 94(5), 2621–4. doi:10.3168/jds.2010-3893.
Al-Qaisi, M., Horst, E. A., Kvidera, S. K., Mayorga, E. J., Timms, L. L. and Baumgard, L. H. 2019. Technical note: Developing a heat stress model in dairy cows using an electric heat blanket. *J. Dairy Sci.* 102(1), 684–9. doi:10.3168/jds.2018-15128.
Bastin, C., Berry, D. P., Soyeurt, H. and Gengler, N. 2012. Genetic correlations of days open with production traits and contents in milk of major fatty acids predicted by mid-infrared spectrometry. *J. Dairy Sci.* 95(10), 6113–21. doi:10.3168/jds.2012-5361.
Beja-Pereira, A., Caramelli, D., Lalueza-Fox, C., Vernesi, C., Ferrand, N., Casoli, A., Goyache, F., Royo, L. J., Conti, S., Lari, M., et al. 2006. The origin of European cattle: evidence from modern and ancient DNA. *Proc. Natl. Acad. Sci. U.S.A.* 103(21), 8113–8. doi:10.1073/pnas.0509210103.
Berman, A. 2005. Estimates of heat stress relief needs for Holstein dairy cows. *J. Anim. Sci.* 83(6), 1377–84. doi:10.2527/2005.8361377x.
Berman, A. 2011. Invited review: Are adaptations present to support dairy cattle productivity in warm climates? *J. Dairy Sci.* 94(5), 2147–58. doi:10.3168/jds.2010-3962.
Bernabucci, U., Biffani, S., Buggiotti, L., Vitali, A., Lacetera, N. and Nardone, A. 2014. The effects of heat stress in Italian Holstein dairy cattle. *J. Dairy Sci.* 97(1), 471–86. doi:10.3168/jds.2013-6611.
Bohmanova, J., Misztal, I., Tsuruta, S., Norman, H. D. and Lawlor, T. J. 2005. National genetic evaluation of milk yield for heat tolerance of United States Holsteins. *Interbull Bulletin* 33, 160–2.
Bohmanova, J., Misztal, I. and Cole, J. B. 2007. Temperature-humidity indices as indicators of milk production losses due to heat stress. *J. Dairy Sci.* 90(4), 1947–56. doi:10.3168/jds.2006-513.
Bradley, D. G., MacHugh, D. E., Cunningham, P. and Loftus, R. T. 1996. Mitochondrial diversity and the origins of African and European cattle. *Proc. Natl Acad. Sci. U.S.A.* 93(10), 5131–5. doi:10.1073/pnas.93.10.5131.
Brügemann, K., Gernand, E., von Borstel, U. U. and König, S. 2011. Genetic analyses of protein yield in dairy cows applying random regression models with time-dependent and temperature × humidity-dependent covariates. *J. Dairy Sci.* 94(8), 4129–39. doi:10.3168/jds.2010-4063.

© Burleigh Dodds Science Publishing Limited, 2020. All rights reserved.

Calus, M. P. L., Windig, J. J. and Veerkamp, R. F. 2005. Associations among descriptors of herd management and phenotypic and genetic levels of health and fertility. *J. Dairy Sci.* 88(6), 2178–89. doi:10.3168/jds.S0022-0302(05)72893-9.

Carabaño, M. J., Bachagha, K., Ramón, M. and Díaz, C. 2014. Modeling heat stress effect on Holstein cows under hot and dry conditions: selection tools. *J. Dairy Sci.* 97(12), 7889–904. doi:10.3168/jds.2014-8023.

Carabaño, M. J., Logar, B., Bormann, J., Minet, J., Vanrobays, M. L., Díaz, C., Tychon, B., Gengler, N. and Hammami, H. 2016. Modeling heat stress under different environmental conditions. *J. Dairy Sci.* 99(5), 3798–814. doi:10.3168/jds.2015-10212.

Carabaño, M. J., Ramón, M., Díaz, C., Molina, A., Pérez-Guzmán, M. D. and Serradilla, J. M. 2017. Breeding and Genetics Symposium: breeding for resilience to heat stress effects in dairy ruminants. A comprehensive review. *J. Anim. Sci.* 95(4), 1813–26. doi:10.2527/jas.2016.1114.

Carlson, D. F., Lancto, C. A., Zang, B., Kim, E. S., Walton, M., Oldeschulte, D., Seabury, C., Sonstegard, T. S. and Fahrenkrug, S. C. 2016. Production of hornless dairy cattle from genome-edited cell lines. *Nat. Biotechnol.* 34(5), 479–81. doi:10.1038/nbt.3560.

Collier, R. J., Dahl, G. E. and VanBaale, M. J. 2006. Major advances associated with environmental effects on dairy cattle. *J. Dairy Sci.* 89(4), 1244–53. doi:10.3168/jds.S0022-0302(06)72193-2.

Collier, R. J., Collier, J. L., Rhoads, R. P. and Baumgard, L. H. 2008. Invited review: Genes involved in the Bovine heat stress response. *J. Dairy Sci.* 91(2), 445–54. doi:10.3168/jds.2007-0540.

Conte, G., Ciampolini, R., Cassandro, M., Lasagna, E., Calamari, L., Bernabucci, U. and Abeni, F. 2018. Feeding and nutrition management of heat-stressed dairy ruminants. *Ital. J. Anim. Sci.* 17(3), 604–20. doi:10.1080/1828051X.2017.1404944.

DairyBio. 2018. Annual progress report 2016–17. ISSN 2209-1424).

Davis, S. R., Spelman, R. J. and Littlejohn, M. D. 2017. Breeding and Genetics Symposium: breeding heat tolerant dairy cattle: the case for introgression of the 'slick' prolactin receptor variant into dairy breeds. *J. Anim. Sci.* 95(4), 1788–800. doi:10.2527/jas.2016.0956.

de Jong, G. and Bijma, P. 2002. Selection and phenotypic plasticity in evolutionary biology and animal breeding. *Livest. Prod. Sci.* 78(3), 195–214. doi:10.1016/S0301-6226(02)00096-9.

Dikmen, S. and Hansen, P. J. 2009. Is the temperature-humidity index the best indicator of heat stress in lactating dairy cows in a subtropical environment? *J. Dairy Sci.* 92(1), 109–16. doi:10.3168/jds.2008-1370.

Dikmen, S., Khan, F. A., Huson, H. J., Sonstegard, T. S., Moss, J. I., Dahl, G. E. and Hansen, P. J. 2014. The SLICK hair locus derived from Senepol cattle confers thermotolerance to intensively managed lactating Holstein cows. *J. Dairy Sci.* 97(9), 5508–20. doi:10.3168/jds.2014-8087.

Dowling, D. F. 1955. The thickness of cattle skin. *Aust. J. Agric. Res.* 6(5), 776–85. doi:10.1071/AR9550776.

Egger-Danner, C., Cole, J. B., Pryce, J. E., Gengler, N., Heringstad, B., Bradley, A. and Stock, K. F. 2015. Invited review: Overview of new traits and phenotyping strategies in dairy cattle with a focus on functional traits. *Animal* 9(2), 191–207. doi:10.1017/S1751731114002614.

Finch, V. A. 1985. Comparison of non-evaporative heat transfer in different cattle breeds. *Aust. J. Agric. Res.* 36(3), 497–508. doi:10.1071/AR9850497.

Finocchiaro, R., van Kaam, J. B., Portolano, B. and Misztal, I. 2005. Effect of heat stress on production of Mediterranean dairy sheep. *J. Dairy Sci.* 88(5), 1855–64. doi:10.3168/jds.S0022-0302(05)72860-5.

Fodor, N., Foskolos, A., Topp, C. F. E., Moorby, J. M., Pásztor, L. and Foyer, C. H. 2018. Spatially explicit estimation of heat stress-related impacts of climate change on the milk production of dairy cows in the United Kingdom. *PLoS ONE* 13(5), e0197076. doi:10.1371/journal.pone.0197076.

Freitas, M. S., Misztal, I., Bohmanova, J. and West, J. 2006. Utility of on- and off-farm weather records for studies in genetics of heat tolerance. *Livest. Sci.* 105(1–3), 223–8. doi:10.1016/j.livsci.2006.06.011.

© Burleigh Dodds Science Publishing Limited, 2020. All rights reserved.

Galukande, E., Mulindwa, H., Wurzinger, M., Roschinsky, R., Mwai, A. O. and Sölkner, J. 2013. Cross-breeding cattle for milk production in the tropics: achievements, challenges and opportunities. *Anim. Genet. Resour.* 52, 111–25.

Gantner, V., Bobić, T., Gregić, M., Gantner, R., Kuterovac, K. and Potočnik, K. 2017. The differences in heat stress resistance due to dairy cattle breed. *Mljekarstvo* 67, 112–22.

Garner, J. B., Douglas, M., Williams, R. S. O., Wales, W. J., Marett, L. C., Nguyen, T. T. T., Reich, C. M. and Hayes, B. J. 2016. Genomic selection improves heat tolerance in dairy cattle. *Nat. Sci. Rep.* 6, 1–8.

Garner, J. B., Douglas, M., Williams, S. R. O., Wales, W. J., Marett, L. C., DiGiacomo, K., Leury, B. J. and Hayes, B. J. 2017. Responses of dairy cows to short-term heat stress in controlled-climate chambers. *Anim. Prod. Sci.* 57(7), 1233–41. doi:10.1071/AN16472.

Gengler, N., Berry, D. P. and Bastin, C. 2013. Use of automated systems for recording of direct and indirect data with special emphasis on the use of MIR milk spectra (OptiMIR project). Challenges and benefits of health data recording in the context of food chain quality, management and breeding. In: *Proceedings of the ICAR Health Data Conference*, 30–31 May 2013. ICAR Technical Series no. 17, Aarhus, Denmark, pp. 55–61.

Glass, E. J., Preston, P. M., Springbett, A., Craigmile, S., Kirvar, E., Wilkie, G. and Brown, C. G. 2005. *Bos taurus* and *Bos indicus* (Sahiwal) calves respond differently to infection with *Theileria annulata* and produce markedly different levels of acute phase proteins. *Int. J. Parasitol.* 35(3), 337–47. doi:10.1016/j.ijpara.2004.12.006.

Goddard, M. E., Kemper, K. E., MacLeod, I. M., Chamberlain, A. J. and Hayes, B. J. 2016. Genetics of complex traits: prediction of phenotype, identification of causal polymorphisms and genetic architecture. *Proc. Biol. Sci.* 283(1835). doi:10.1098/rspb.2016.0569.

Gonzalez-Rivas, P. A., Sullivan, M., Cottrell, J. J., Leury, B. J., Gaughan, J. B. and Dunshea, F. R. 2018. Effect of feeding slowly fermentable grains on productive variables and amelioration of heat stress in lactating dairy cows in a sub-tropical summer. *Trop. Anim. Health Prod.* 50(8), 1763–9. doi:10.1007/s11250-018-1616-5.

Haile-Mariam, M., Bowman, P. J. and Pryce, J. E. 2013. Genetic analyses of fertility and predictor traits in Holstein herds with low and high mean calving intervals and in Jersey herds. *J. Dairy Sci.* 96(1), 655–67. doi:10.3168/jds.2012-5671.

Hammami, H., Bormann, J., M'hamdi, N., Montaldo, H. H. and Gengler, N. 2013. Evaluation of heat stress effects on production traits and somatic cell score of Holsteins in a temperate environment. *J. Dairy Sci.* 96(3), 1844–55. doi:10.3168/jds.2012-5947.

Hammami, H., Vandenplas, J., Vanrobays, M. L., Rekik, B., Bastin, C. and Gengler, N. 2015. Genetic analysis of heat stress effects on yield traits, udder health, and fatty acids of Walloon Holstein cows. *J. Dairy Sci.* 98(7), 4956–68. doi:10.3168/jds.2014-9148.

Hammond, A. C. and Olson, T. A. 1994. Rectal temperature and grazing time in selected beef cattle breeds under tropical summer conditions in subtropical Florida. *J. Trop. Agric. (Trinidad)* 71, 128–34.

Hammond, A. C., Olson, T. A., Chase, C. C., Jr., Bowers, E. J., Randel, R. D., Murphy, C. N., Vogt, D. W. and Tewolde, A. 1996. Heat tolerance in two tropically adapted *Bos taurus* breeds, Senepol and Romosinuano, compared with Brahman, Angus, and Hereford cattle in Florida. *J. Anim. Sci.* 74(2), 295–303. doi:10.2527/1996.742295x.

Hammond, A. C., Chase, C. C., Jr., Bowers, E. J., Olson, T. A. and Randel, R. D. 1998. Heat tolerance in Tuli-, Senepol-, and Brahman-sired F1 Angus heifers in Florida. *J. Anim. Sci.* 76(6), 1568–77. doi:10.2527/1998.7661568x.

Hansen, P. J. 2004. Physiological and cellular adaptations of zebu cattle to thermal stress. *Anim. Reprod. Sci.* 82–83, 349–60. doi:10.1016/j.anireprosci.2004.04.011.

Hansen, P. J. and Aréchiga, C. F. 1999. Strategies for managing reproduction in the heat-stressed dairy cows. *J. Anim. Sci.* 77, 36–50. doi:10.2527/1997.77suppl_236x.

Hayes, B. J., Carrick, M., Bowman, P. and Goddard, M. E. 2003. Genotype × environment interaction for milk production of daughters of Australian dairy sires from test-day records. *J. Dairy Sci.* 86(11), 3736–44. doi:10.3168/jds.S0022-0302(03)73980-0.

© Burleigh Dodds Science Publishing Limited, 2020. All rights reserved.

Hayes, B. J., Bowman, P. J., Chamberlain, A. J., Savin, K., van Tassell, C. P., Sonstegard, T. S. and Goddard, M. E. 2009a. A validated genome wide association study to breed cattle adapted to an environment altered by climate change. PLoS ONE 4(8), 1–8. doi:10.1371/journal.pone.0006676.

Hayes, B. J., Bowman, P. J., Chamberlain, A. J., Verbyla, K. and Goddard, M. E. 2009b. Accuracy of genomic breeding values in multi-breed dairy cattle populations. Genet. Sel. Evol. 41, 1–9.

Hayes, B. J., Daetwyler, H. D. and Goddard, M. E. 2016. Models for genome × environment interaction: examples in livestock. Crop Sci. 56(5), 2251–9. doi:10.2135/cropsci2015.07.0451.

Hiendleder, S., Lewalski, H. and Janke, A. 2008. Complete mitochondrial genomes of Bos taurus and Bos indicus provide new insights into intra-species variation, taxonomy and domestication. Cytogenet. Genome Res. 120(1–2), 150–6. doi:10.1159/000118756.

Honig, H., Ofer, L., Kaim, M., Jacobi, S., Shinder, D. and Gershon, E. 2016. The effect of cooling management on blood flow to the dominant follicle and estrous cycle length at heat stress. Theriogenology 86(2), 626–34. doi:10.1016/j.theriogenology.2016.02.017.

IPCC. 2014. Climate Change 2014: Synthesis Report. Contribution of Working Groups I, II and III to the Fifth Assessment Report of the Intergovernmental Panel on Climate Change. IPCC, Geneva, Switzerland, 151pp.

Johnston, J. E., Hamblin, F. B. and Schrader, G. T. 1958. Factors concerned in the comparative heat tolerance of Jersey, Holstein, and Red Sindhi-Holstein (F1) cattle. J. Anim. Sci. 17(2), 473–9. doi:10.2527/jas1958.172473x.

Kadzere, C. T., Murphy, M. R., Silanikove, N. and Maltz, E. 2002. Heat stress in lactating dairy cows: a review. Livest. Prod. Sci. 77(1), 59–91. doi:10.1016/S0301-6226(01)00330-X.

Kandel, P. B., Vanrobays, M. L., Vanlierde, A., Dehareng, F., Froidmont, E., Dardenne, P., Lewis, E., Buckley, F., Deighton, M. H., McParland, S., et al. 2013. Genetic parameters for methane emission predicted from milk mid-infrared spectra in dairy cows. Adv. Anim. Biosci. 4, 279.

Kolmodin, R., Strandberg, E., Madsen, P., Jensen, J. and Jorjani, H. 2002. Genotype by environment interaction in Nordic dairy cattle studied using reaction norms. Acta Agric. Scand. A Anim. Sci. 52(1), 11–24. doi:10.1080/09064700252806380.

Koltes, J. E., Koltes, D. A., Mote, B. E., Tucker, J. and Hubbell, I. I. I. D. S. 2018. Automated collection of heat stress data in livestock: new technologies and opportunities. Transl. Anim. Sci. 2, 319–23.

Li, X., Wang, S., Huang, J., Li, L., Zhang, Q. and Ding, X. 2014. Improving the accuracy of genomic prediction in Chinese Holstein cattle by using one-step blending. Genet. Sel. Evol. 46, 66–. doi:10.1186/s12711-014-0066-4.

Littlejohn, M. D., Henty, K. M., Tiplady, K., Johnson, T., Harland, C., Lopdell, T., Sherlock, R. G., Li, W., Lukefahr, S. D., Shanks, B. C., et al. 2014. Functionally reciprocal mutations of the prolactin signalling pathway define hairy and slick cattle. Nat. Commun. 5, 5861. doi:10.1038/ncomms6861.

Liu, X., Wang, Y., Guo, W., Chang, B., Liu, J., Guo, Z., Quan, F. and Zhang, Y. 2013. Zinc-finger nickase-mediated insertion of the lysostaphin gene into the beta-casein locus in cloned cows. Nat. Commun. 4, 2565. doi:10.1038/ncomms3565.

Liu, X., Wang, Y., Tian, Y., Yu, Y., Gao, M., Hu, G., Su, F., Pan, S., Luo, Y., Guo, Z., et al. 2014. Generation of mastitis resistance in cows by targeting human lysozyme gene to beta-casein locus using zinc-finger nucleases. Proc. Biol. Sci. 281(1780), 20133368. doi:10.1098/rspb.2013.3368.

Luo, J., Song, Z., Yu, S., Cui, D., Wang, B., Ding, F., Li, S., Dai, Y. and Li, N. 2014. Efficient generation of myostatin (MSTN) biallelic mutations in cattle using zinc finger nucleases. PLoS ONE 9(4), e95225. doi:10.1371/journal.pone.0095225.

Macciotta, N. P. P., Biffani, S., Bernabucci, U., Lacetera, N., Vitali, A., Ajmone-Marsan, P. and Nardone, A. 2017. Derivation and genome-wide association study of a principal component-based measure of heat tolerance in dairy cattle. J. Dairy Sci. 100(6), 4683–97. doi:10.3168/jds.2016-12249.

MacHugh, D. E., Shriver, M. D., Loftus, R. T., Cunningham, P. and Bradley, D. G. 1997. Microsatellite DNA variation and the evolution, domestication and phylogeography of taurine and zebu cattle (Bos taurus and Bos indicus). Genetics 146(3), 1071–86.

© Burleigh Dodds Science Publishing Limited, 2020. All rights reserved.

MacLeod, I. M., Bowman, P. J., Vander Jagt, C. J., Haile-Mariam, M., Kemper, K. E., Chamberlain, A. J., Schrooten, C., Hayes, B. J. and Goddard, M. E. 2016. Exploiting biological priors and sequence variants enhances QTL discovery and genomic prediction of complex traits. *BMC Genomics* 17, 144. doi:10.1186/s12864-016-2443-6.

Madalena, F. E., Teodoro, R. L., Lemos, A. M., Monteiro, J. B. N. and Barbosa, R. T. 1990. Evaluation of strategies for crossbreeding of dairy cattle in brazil. *J. Dairy Sci.* 73(7), 1887–901. doi:10.3168/jds.S0022-0302(90)78869-8.

Masuda, Y., Misztal, I., Tsuruta, S., Legarra, A., Aguilar, I., Lourenco, D. A. L., Fragomeni, B. O. and Lawlor, T. J. 2016. Implementation of genomic recursions in single-step genomic best linear unbiased predictor for US Holsteins with a large number of genotyped animals. *J. Dairy Sci.* 99(3), 1968–74. doi:10.3168/jds.2015-10540.

Meuwissen, T. H., Hayes, B. J. and Goddard, M. E. 2001. Prediction of total genetic value using genome-wide dense marker maps. *Genetics* 157(4), 1819–29.

Meuwissen, T. H., Hayes, B. J. and Goddard, M. E. 2016. Genomic selection: a paradigm shift in animal breeding. *Anim. Front.* 6(1), 6–14. doi:10.2527/af.2016-0002.

Min, L., Cheng, J., Zhao, S., Tian, H., Zhang, Y., Li, S., Yang, H., Zheng, N. and Wang, J. 2016. Plasma-based proteomics reveals immune response, complement and coagulation cascades pathway shifts in heat-stressed lactating dairy cows. *J. Proteom.* 146, 99–108. doi:10.1016/j.jprot.2016.06.008.

Min, L., Zhao, S., Tian, H., Zhou, X., Zhang, Y., Li, S., Yang, H., Zheng, N. and Wang, J. 2017. Metabolic responses and 'omics' technologies for elucidating the effects of heat stress in dairy cows. *Int. J. Biometeorol.* 61(6), 1149–58. doi:10.1007/s00484-016-1283-z.

Misztal, I., Bohmanova, J., Freitas, M., Tsuruta, S., Norman, H. D. and Lawlor, T. J. 2006. Issues in genetic evaluation of dairy cattle for heat stress. In: *Proceedings Therapeutic 8th World Congress on Genetics Applied to Livestock Production*, 13–18 August 2006, Belo Horizonte, MG, Brazil.

Misztal, I., Legarra, A. and Aguilar, I. 2009. Computing procedures for genetic evaluation including phenotypic, full pedigree, and genomic information. *J. Dairy Sci.* 92(9), 4648–55. doi:10.3168/jds.2009-2064.

Misztal, I., Legarra, A. and Aguilar, I. 2014. Using recursion to compute the inverse of the genomic relationship matrix. *J. Dairy Sci.* 97(6), 3943–52. doi:10.3168/jds.2013-7752.

Nay, T. and Heyman, R. H. 1956. Sweat glands in zebu (*Bos indicus* L.) and European (*B. taurus* L.) cattle. I. Size of individual glands, the denseness of their population, and their depth below the skin surface. *Aust. J. Agric. Res.* 7(5), 482–94. doi:10.1071/AR9560482.

Nguyen, T. T. T., Bowman, P. J., Haile-Mariam, M., Pryce, J. E. and Hayes, B. J. 2016. Genomic selection for heat tolerance in Australian dairy cattle. *J. Dairy Sci.* 99(4), 2849–62. doi:10.3168/jds.2015-9685.

Nguyen, T. T. T., Bowman, P. J., Haile-Mariam, M., Nieuwhof, G. J., Hayes, B. J. and Pryce, J. E. 2017a. Short communication: Implementation of a breeding value for heat tolerance in Australian dairy cattle. *J. Dairy Sci.* 100(9), 7362–7. doi:10.3168/jds.2017-12898.

Nguyen, T. T. T., Garner, J. B., Bowman, P. J., Haile-Mariam, M., Pryce, J. E. and Hayes, B. J. 2017b. Breeding value for heat tolerance in Australian dairy cattle: a technical platform for implementation. Herd 17. DataGene, Bendigo, Australia, pp. 35–9.

Olson, T. A., Lucena, C., Chase, C. C. and Hammond, A. C. 2003. Evidence of a major gene influencing hair length and heat tolerance in *Bos taurus* cattle. *J. Anim. Sci.* 81(1), 80–90. doi:10.2527/2003.81180x.

Pan, Y. S. 1963. Quantitative and morphological variation of sweat glands, skin thickness, and skin shrinkage over various body regions of Sahiwal Zebu and Jersey cattle. *Aust. J. Agric. Res.* 14(3), 424–37. doi:10.1071/AR9630424.

Proudfoot, C., Carlson, D. F., Huddart, R., Long, C. R., Pryor, J. H., King, T. J., Lillico, S. G., Mileham, A. J., McLaren, D. G., Whitelaw, C. B., et al. 2015. Genome edited sheep and cattle. *Transgenic Res.* 24(1), 147–53. doi:10.1007/s11248-014-9832-x.

© Burleigh Dodds Science Publishing Limited, 2020. All rights reserved.

Pryce, J. E., Douglas, P., Reich, C. M., Chamberlain, A. C., Bowman, P. J., Nguyen, T. T. T., Mason, B. A., Prowse-Wilkins, C. P., Nieuwhof, G. J., Hancock, T., et al. 2017. Reliabilities of Australian dairy genomic breeding values increase through the addition of genotyped females with excellent phenotypes. In: *Proceedings of Association of the Advancement of Animal Breeding and Genetics*. AAABG, Townsville, pp. 133–6.

Pryce, J. E., Nguyen, T. T. T., Axford, M., Nieuwhof, G. and Shaffer, M. 2018. Symposium review: Building a better cow—The Australian experience and future perspectives. *J. Dairy Sci.* 101(4), 3702–13. doi:10.3168/jds.2017-13377.

Purwanto, B. P., Abo, Y., Sakamoto, R., Furumoto, F. and Yamamoto, S. 1990. Diurnal patterns of heat production and heart rate under thermoneutral conditions in Holstein Friesian cows differing in milk production. *J. Agric. Sci.* 114(2), 139–42. doi:10.1017/S0021859600072117.

Ravagnolo, O. and Misztal, I. 2000. Genetic component of heat stress in dairy cattle, parameter estimation. *J. Dairy Sci.* 83(9), 2126–30. doi:10.3168/jds.S0022-0302(00)75095-8.

Renaudeau, D., Collin, A., Yahav, S., de Basilio, V., Gourdine, J. L. and Collier, R. J. 2012. Adaptation to hot climate and strategies to alleviate heat stress in livestock production. *Animal* 6(5), 707–28. doi:10.1017/S1751731111002448.

Roberts, T., Chapinal, N., Leblanc, S. J., Kelton, D. F., Dubuc, J. and Duffield, T. F. 2012. Metabolic parameters in transition cows as indicators for early-lactation culling risk. *J. Dairy Sci.* 95(6), 3057–63. doi:10.3168/jds.2011-4937.

Sánchez, J. P., Misztal, I., Aguilar, I., Zumbach, B. and Rekaya, R. 2009. Genetic determination of the onset of heat stress on daily milk production in the US Holstein cattle. *J. Dairy Sci.* 92(8), 4035–45. doi:10.3168/jds.2008-1626.

Seif, S. M., Johnson, H. D. and Lippincott, A. C. 1979. The effects of heat exposure (31°C) on Zebu and Scottish Highland cattle. *Int. J. Biometeorol.* 23(1), 9–14. doi:10.1007/BF01553372.

Silanikove, N. 1992. Effects of water scarcity and hot environment on appetite and digestion in ruminants: a review. *Livest. Prod. Sci.* 30(3), 175–94. doi:10.1016/S0301-6226(06)80009-6.

Sonna, L. A., Fujita, J., Gaffin, S. L. and Lilly, C. M. 2002. Invited review: Effects of heat and cold stress on mammalian gene expression. *J. Appl. Physiol.* 92(4), 1725–42. doi:10.1152/japplphysiol.01143.2001.

St-Pierre, N. R., Cobanov, B. and Schnitkey, G. 2003. Economic losses from heat stress by US livestock industries. *J. Dairy Sci.* 86, 52–9.

Strandén, I., Matilainen, K., Aamand, G. P. and Mantysaari, E. A. 2017. Solving efficiently large single-step genomic best linear unbiased prediction models. *J. Anim. Breed. Genet.* 134(3), 264–74. doi:10.1111/jbg.12257.

Tan, W., Carlson, D. F., Lancto, C. A., Garbe, J. R., Webster, D. A., Hackett, P. B. and Fahrenkrug, S. C. 2013. Efficient nonmeiotic allele introgression in livestock using custom endonucleases. *Proc. Natl Acad. Sci. U.S.A.* 110(41), 16526–31. doi:10.1073/pnas.1310478110.

Tao, S. and Dahl, G. E. 2013. Invited review: Heat stress effects during late gestation on dry cows and their calves. *J. Dairy Sci.* 96(7), 4079–93. doi:10.3168/jds.2012-6278.

Tian, H., Wang, W., Zheng, N., Cheng, J., Li, S., Zhang, Y. and Wang, J. 2015. Identification of diagnostic biomarkers and metabolic pathway shifts of heat-stressed lactating dairy cows. *J. Proteom.* 125, 17–28. doi:10.1016/j.jprot.2015.04.014.

Tian, H., Zheng, N., Wang, W., Cheng, J., Li, S., Zhang, Y. and Wang, J. 2016. Integrated metabolomics study of the milk of heat-stressed lactating dairy cows. *Sci. Rep.* 6, 24208. doi:10.1038/srep24208.

Utech, K. B. and Wharton, R. H. 1982. Breeding for resistance to Boophilus microplus in Australian Illawarra Shorthorn and Brahman × Australian Illawarra Shorthorn cattle. *Aust. Vet. J.* 58(2), 41–6. doi:10.1111/j.1751-0813.1982.tb02684.x.

van der Drift, S. G. A., Jorritsma, R., Schonewille, J. T., Knijn, H. M. and Stegeman, J. A. 2012. Routine detection of hyperketonemia in dairy cows using fourier transform infrared spectroscopy analysis of beta-hydroxybutyrate and acetone in milk in combination with test-day information. *J. Dairy Sci.* 95(9), 4886–98. doi:10.3168/jds.2011-4417.

© Burleigh Dodds Science Publishing Limited, 2020. All rights reserved.

Vanlierde, A., Vanrobays, M.-L., Gengler, N., Dardenne, P., Froidmont, E., Soyeurt, H., McParland, S., Lewis, E., Deighton, M. H., Mathot, M., et al. 2016. Milk mid-infrared spectra enable prediction of lactation-stage-dependent methane emissions of dairy cattle within routine population-scale milk recording schemes. *Anim. Prod. Sci.* 56(3), 258–64. doi:10.1071/AN15590.

Vermunt, J. J. and Tranter, B. P. 2011. Heat stress in dairy cattle – a review, and some of the potential risks associated with the nutritional management of this condition. In: *Proceedings of Annual Conference of the Australian Veterinary Association*. Queensland Division, Townsville, QLD, Australia, pp. 212–21.

Wambura, P. N., Gwakisa, P. S., Silayo, R. S. and Rugaimukamu, E. A. 1998. Breed-associated resistance to tick infestation in *Bos indicus* and their crosses with *Bos taurus*. *Vet. Parasitol.* 77(1), 63–70. doi:10.1016/S0304-4017(97)00229-X.

West, J. W. 2003. Effects of heat-stress on production in dairy cattle. *J. Dairy Sci.* 86(6), 2131–44. doi:10.3168/jds.S0022-0302(03)73803-X.

Wu, H., Wang, Y., Zhang, Y., Yang, M., Lv, J., Liu, J. and Zhang, Y. 2015. TALE nickase-mediated SP110 knockin endows cattle with increased resistance to tuberculosis. *Proc. Natl Acad. Sci. U.S.A.* 112(13), E1530–9. doi:10.1073/pnas.1421587112.

Yousef, M. K. 1985. *Stress Physiology in Livestock*. CRC Press, Boca Raton, FL.

Yu, S., Luo, J., Song, Z., Ding, F., Dai, Y. and Li, N. 2011. Highly efficient modification of *beta-lactoglobulin* (*BLG*) gene via zinc-finger nucleases in cattle. *Cell Res.* 21(11), 1638–40. doi:10.1038/cr.2011.153.

Zimbelman, R. B., Rhoads, R. P., Rhoads, M. L., Duff, G. C., Baumgard, L. H. and Collier, R. J. 2009. A re-evaluation of the impact of temperature humidity index (THI) and black globe humidity index (BGHI) on milk production in high producing dairy cows. In: *Southwest Nutrition Management Conference Tempe*, ARPAS, Savoy, IL, pp. 158–68.

© Burleigh Dodds Science Publishing Limited, 2020. All rights reserved.

Chapter 2

Genetic selection for dairy cow welfare and resilience to climate change

Jennie E. Pryce, Agriculture Victoria and La Trobe University, Australia; and Yvette de Haas, Wageningen UR, The Netherlands

1 Introduction

Rates of genetic gain in dairy cows are impressive, especially for milk production traits that drive profitability (VanRaden 2004). However, narrow breeding goals focused on milk production traits are detrimental to reproductive performance and health of cows (Rauw et al. 1998), and consequently there has been pressure to develop breeding values to enable selection that balances both production and non-production traits. Although, the rationale to extend breeding goals initially focused entirely on the impact of the new breeding value to farmer profitability, breeding goals are now becoming more complex in order to meet challenges set by consumers and society (Boichard and Brochard 2012; Martin-Collado et al. 2015). For example, the growing human population places more pressure on the available limited resources; global changes may mean hotter drier conditions to manage livestock and there is also increased consumer awareness about animal welfare and farming conditions. To accommodate this requirement, over recent decades there has been a rapid expansion of the number of breeding values that are available for farmers to select on. Almost without exception these breeding values rely on large amounts of field data that are freely available through current recording systems, such as milk production, calving records, insemination dates, pregnancy test outcomes, health records and culling dates.

http://dx.doi.org/10.19103/AS.2016.0006.04
© Burleigh Dodds Science Publishing Limited, 2017. All rights reserved.

Box 1 Definitions

A **breeding value**, which is the additive genetic merit of an animal, is predicted from the animal's own performance, the performance of its relatives and more recently using its genomic information through genetic markers. Breeding values are usually estimated using methodology known as best linear unbiased prediction, which is a statistical method that estimates management, feeding and environmental effects simultaneously with estimating breeding values.

Breeding goals, or breeding objectives, are characteristics which the selection is intending to improve, for example profitability, mastitis resistance and so on.

A **selection index** is an overall score of genetic merit that combines information (usually breeding values) on several measurable traits, where each trait's breeding value is weighted using a value that is generally associated with its contribution to the overall breeding objective, for example, profit. Examples of national dairy indices include Economic Breeding Index (EBI) in Ireland, Balanced Performance Index (BPI) in Australia, Breeding Worth (BW) in New Zealand, Profitable Lifetime Index (£PLI) in the United Kingdom, Net Merit (NM) in the United States and so on.

Although breeding values calculated from field data will still be important into the future, the rise of genomic selection as a way of providing breeding values to genotyped animals that do not have phenotypic information, means we are likely to see increased use of reference data (to estimate breeding values). Furthermore, it is likely that the data required will be from research herds or commercial herds and have much more depth of phenotyping than has been possible before. Particularly, this field has huge opportunities for geneticists to work with veterinarians and animal scientists to develop breeding values for new traits that are only measurable on comparatively small genotyped populations.

In this chapter, we will discuss the principles behind breeding goals and multi-trait selection, examine the consequences of selecting for milk production traits and then focus on tangible results in multi-trait selection including developments in breeding values that should lead to improved animal welfare.

The definitions of some key terms in this discussion are provided in Box 1.

2 Selection indices

The derivation of a selection index starts with the definition of an overall breeding objective; usually, this is profitability. As early as 1960, it was recognised that selecting animals objectively using an economic framework was likely to result in greater farm profit (Lush 1960).

The purpose of the selection index method is to combine information from different sources in such a way that the breeding objective is optimised. Index traits, or selection criteria, are where measurements exist, that is, for which breeding values can be predicted. Selection index traits in dairying typically include traits such as milk production, longevity, fertility, health traits, weight (to account for maintenance costs) and conformation associated with performance, for example, udder and legs. There is usually an overlap between selection criteria traits and breeding objective traits, because in order to improve a trait it is always best to measure it. However, some traits may be harder to measure, for example, mastitis resistance. Then, it is often easier to use selection criteria that are easier

© Burleigh Dodds Science Publishing Limited, 2017. All rights reserved.

to measure, for example, somatic cell count (SCC) to predict mastitis resistance, as the genetic correlation is reasonably high at 0.7 (Mrode and Swanson 1996), implying that selection to reduce cell counts will also reduce the occurrence of mastitis.

The next stage in developing a selection index is to calculate economic values for each trait, generally with a bioeconomic model, where the economic value is the increase in revenue from a unit change of a trait while everything else is held constant. Then, selection index theory (Hazel 1943) is commonly used to calculate the most appropriate index weights and responses to selection for a set of traits given the genetic and phenotypic (co) variances and the economic values of traits in the index. The resulting selection index is the sum of n estimated breeding values (EBV_i) for each trait multiplied by their respective index weights b_i

$$\text{Index} = b_1 EBV_1 + b_2 EBV_2 + \ldots\ldots + b_n EBV_n$$

For many countries, the weights used in selection indices are calculated using solely economic parameters. However, there is growing awareness that to meet future economic, societal and environmental requirements the optimal weighting on breeding goal traits needs to be reconsidered. Examples of devising non-market values for these traits are discussed by Nielsen et al. (2005) and include consumer willingness to pay for aspects of traits that have perceived societal or animal welfare value. Another approach, described by Martin-Collado et al. (2015) uses '1000 minds' methodology to add objectivity to perceived non-market values through a survey, where questions on perceived values are assessed through a series of comparisons that are of similar actual value. The idea being that if opinions are canvassed from many farmers (hence the '1000 minds' name), the comparative value of a trait to groups of farmers with similar philosophies can be quantified.

Farmer preferences for the improvement of key traits in Australia were assessed by the degree that perceived value deviated most from the trait's actual economic value (Martin-Collado et al. 2015) and the ranking was as follows: (1) mastitis, (2) longevity, (3) fertility, (4) mammary system, (5) lameness, (6) protein yield, (7) type, (8) feed efficiency, (9) calving ease, (10) temperament, (11) lactation persistence and (12) liveweight. Although farmer preferences were the focus of the research, it was clear that animal welfare (with three traits ranked in the top five associated with health) and improving the functional ability of dairy cows was at the forefront of farmers' preferences for future generations of their herds. Accordingly, extra weight was applied to traits that were rated highly to generate three indices that were released in 2015 by the Australian Dairy Herd Improvement Scheme for farmers to use: the Balanced Performance Index (BPI), which is focused on profitability and designed to be in line with farmer preferences; the Health Weighted Index, which has additional selection emphasis on health, fertility and survival traits; and the Type Weighted Index, which places additional selection pressure on conformation traits.

3 Selection for milk production, energy balance and fertility

3.1 Selection for milk production traits

In the late twentieth century, it was acknowledged that selection for milk production traits (milk quality and quantity) had been very successful (VanRaden 2004) and was largely

© Burleigh Dodds Science Publishing Limited, 2017. All rights reserved.

attributable to world domination of the highly productive Holstein breed. Average per cow production in the United States has increased by 6763 litres between 1957 and 2007 (https://www.cdcb.us/eval/summary/trend.cfm). Around 54% of this improvement is due to genetics through the use of artificial insemination and selection based on progeny testing and, more recently, on genomic selection and the widespread use of elite sires through international marketing of bull semen.

By the 1980s and 1990s evidence was starting to build that selection for milk production traits had led to unwanted consequences in other traits of importance, notably unfavourable genetic correlations between fertility and milk production traits (Pryce and Veerkamp 1999; Berry et al. 2014), but there was also evidence that other traits associated with health and animal welfare were also starting to deteriorate (Uribe et al. 1995).

Important thinking at the time led to the development of resource allocation theory, where the resources that an animal has are limited, so if output (milk production) increases, other functions, such as fertility and immune function would suffer (Beilharz et al. 1993). Following on from this, Rauw et al. (1998) concluded that in populations that are genetically driven towards high production welfare would be compromised when resources are limited. At around the same time, research was starting to focus on the effects of metabolic load defined as the 'burden imposed by the synthesis and secretion of milk' (Knight et al. 1999). If metabolic load surpassed a threshold, then it could lead to metabolic stress 'that amount of metabolic load which cannot be sustained, such that some energetic processes, including those that maintain general health must be down regulated' (Knight et al. 1999). The extent of downregulation could be indicative of the degree of metabolic stress. One way in which metabolic load can be quantified is using indicators of perturbations in health and fertility.

3.2 Energy balance

A trait that underpins many health and fertility traits is energy balance (Mallard et al. 2000), which is the difference between energy intake and energy requirements; when intake does not match expenditure cows are described as being in negative energy balance. Energy balance can only be accurately assessed using calorimeter chambers, if feed intake data is available then an approximation of energy input minus energy output can be made (Banos et al. 2005; Friggens et al. 2007a). Several predictors of energy balance have been proposed, including body condition score change (Roche et al. 2009) and fat-to-protein ratio (Friggens et al. 2007a). Body condition score is a subjective measure to determine the body reserves of an animal and it decreases as body reserves are mobilised to compensate negative energy balance. However, the advent of automated weighing and the promising results in semi-automating body condition scoring (Bewley et al. 2008) makes the use of longitudinal records across lactation a practical novel strategy to calculate energy balance (Thorup et al. 2012). Clearly, this is a very important trait from a welfare perspective, as some have argued that excessive mobilisation of body tissue is indicative of starvation (Oltenacu and Broom 2010).

Genetic correlations between yield and energy balance are almost always negative (Veerkamp 1998), implying that selection for yield leads to cows that are more at risk to periods of negative energy balance. There is also ample evidence that this is detrimental to cow health and fertility. Notably, several studies have shown a relationship between energy balance or body condition score and reproductive performance (De Vries and Veerkamp 2000; Pryce et al. 2001; Veerkamp et al. 2003; Friggens et al. 2007a). However, although

© Burleigh Dodds Science Publishing Limited, 2017. All rights reserved.

energy balance may be an intermediary for health and reproductive disorders, the manifestation of disease and reproductive failure is far more complex. Consequently, it is usually more effective to select directly on the traits themselves, generally as part of a multi-trait selection index.

3.3 Fertility

Fertility is a trait of great importance in dairy cattle breeding as lactation is dependent on parturition (i.e. reproduction). It is also a good example of unintended consequences of selection, as dramatic reductions in phenotypic and genetic fertility performance have been universally documented (Berry et al. 2014; Pryce et al. 2014).

Heritability estimates of traditional fertility traits are generally low (<0.1); yet selection for fertility can lead to worthwhile changes because the trait is highly variable (Pryce and Veerkamp 1999; Berry et al. 2014). Pryce et al. (2014) used data collated by the World Holstein Federation to show that phenotypic calving interval appears to have reached a plateau by 2007. Between 1990 and 2000 calving interval increased 'worldwide' by 1.25d/ year phenotypically (Pryce et al. 2014). VanRaden et al. (2004) and Berry et al. (2014) reported that 40 and 64% of the phenotypic decline in performance as measured by calving interval observed in the United States and Ireland respectively, was due to genetics.

Evidence that selecting for female fertility has been successful is starting to emerge. In both Australia and Ireland, phenotypic and genetic trends in reproductive performance (evaluated as calving interval) show that since 2005 reproductive performance is improving (Berry et al. 2014). The same observation has been made in the United States (Fig. 1)

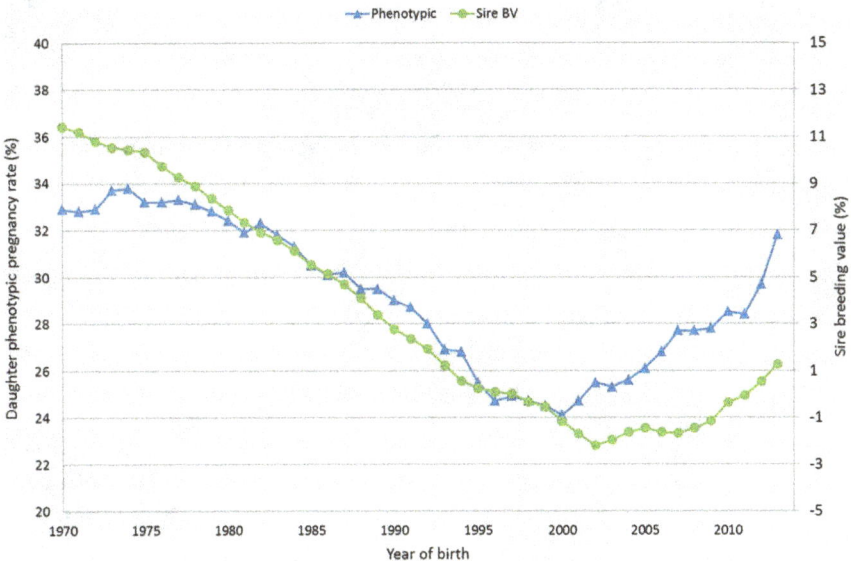

Figure 1 Mean daughter phenotypic pregnancy rate (phenotypic) and sire breeding values (sire BV) for daughter pregnancy rate in US Holstein cows born between 1980 and 2013. Source: https://www.cdcb. us/eval/summary/trend.cfm?R_Menu=HO.d#StartBody April 2016 evaluation.

© Burleigh Dodds Science Publishing Limited, 2017. All rights reserved.

using on average >500 000 observations per year since 1970. In 2014, US phenotypic daughter pregnancy rates appear to have reached the same levels as the early 1980s, as with Ireland and Australia the lowest point for the genetic trends was in the 2000s. Interestingly, improvements in reproductive performance started before the introduction of fertility breeding values in the United States (in 2003), which could have been because of (1) selection on other correlated traits, such as longevity (VanRaden et al. 2004); (2) widespread use of bulls that had positive fertility breeding values (even if this was not known); or (3) more aggressive culling for poor fertility.

As a final note to this section, valuable lessons have been learnt by dairy geneticists and sire analysts in the dangers of narrow breeding goals. However, it appears that tangible improvements in fertility are now being achieved. In addition to sustaining selection on fertility, welfare and disease resistance traits in particular are becoming key areas where breeding values are being developed for future breeding goals.

4 New breeding objectives: health traits

In their review of more than 1600 citations, Kelton et al. (1998) identified eight diseases that had a high economic impact and a large number of cows affected by these diseases (e.g. milk fever, retained placenta, metritis, ketosis, left displaced abomasum, cystic ovarian disease, lameness and clinical mastitis). To be able to reduce disease incidence through breeding, appropriate ways of measuring the traits need to be devised (so-called selection criteria), this can be either measurements of the trait itself (the breeding objective) or traits that are correlated to the breeding objective. In 1988, the first major review of data recording opportunities and consequently breeding strategies to improve production diseases was published (Emanuelson 1988). However, quite a lot has changed since then. Notably, computerised farm recording has led to a large increase in data available on these traits (and their predictors) and consequently studies on genetic parameter and breeding value estimation. In fact, several countries around the world have implemented routine genetic evaluations of health traits using (predominantly) farm-recorded clinical records of disease observations (Egger-Danner et al. 2015). However, there are examples where disease resistance has been selected for indirectly, notably for mastitis.

4.1 Mastitis

4.1.1 Alternative cell count traits

One of the most important diseases in dairy production is mastitis. A common selection criterion, where the breeding goal is to improve mastitis resistance, is selection for reduced SCC. Cell count can be quantified from routinely assessed milk samples and are available to all farmers participating in milk recording to make management decisions. As a by-product of national recording schemes, breeding values for SCC are routinely calculated by many national breeding organisations. The genetic correlation of SCC with mastitis is around 0.7 (Mrode and Swanson 1996), making selection for reduced SCC a convenient way to reduce the incidence of cases of clinical mastitis. The power of using SCC breeding values to reduce the occurrence of clinical mastitis has recently been shown in an Australian study by Abdelsayed et al. (2016; unpublished). In this study, top versus

© Burleigh Dodds Science Publishing Limited, 2017. All rights reserved.

bottom 10% for breeding values of SCC of around 2000 sires with at least 10 cow lactations had a 6.3% difference in the percentage of recorded cases of clinical mastitis (Fig. 2).

However, several studies have shown that selection for directly reducing the number of cases of mastitis is likely to be more effective than relying solely on predictor traits (such as SCC) (Heringstad et al. 2006; Gaddis et al. 2014; Egger-Danner et al. 2015). The Nordic countries have a long history of recording health traits, for example in Norway veterinary treatments had to be registered on an individual basis from 1975 (Heringstad and Østerås 2013), with similar recording being established in Denmark, Finland and Sweden during the 1980s. In addition to the Nordic countries, routine genetic evaluations of mastitis have been in place in Austria and Germany since 2010, and in France and Canada from 2012 (Egger-Danner et al. 2015).

Results from a Norwegian selection experiment have shown that a reduction in the incidence of mastitis is achievable if there is sufficient selection pressure in the breeding objective. Heringstad et al. (2007) demonstrated that after five generations of selection, a 4% difference in clinical mastitis was observed between two lines of Norwegian Red Cattle that were selected for either high protein yield or mastitis resistance, using breeding values calculated with records of clinical mastitis.

As most countries still do not have their health-recording schemes in place yet, several countries have started to look at alternative cell count traits as predictors of clinical (CM) and sub-clinical cases of mastitis (SCM). Alternative SCC traits have been suggested, where test-day records for SCC were either analysed individually (Heuven 1987; Reents et al. 1995) or were described on a lactation level (Detilleux et al. 1997; Schepers et al. 1997; de Haas et al. 2003; Green et al. 2004). Examples of suggested traits are (1) proportions of test-day SCC above or below certain thresholds, (2) directions and rates of change

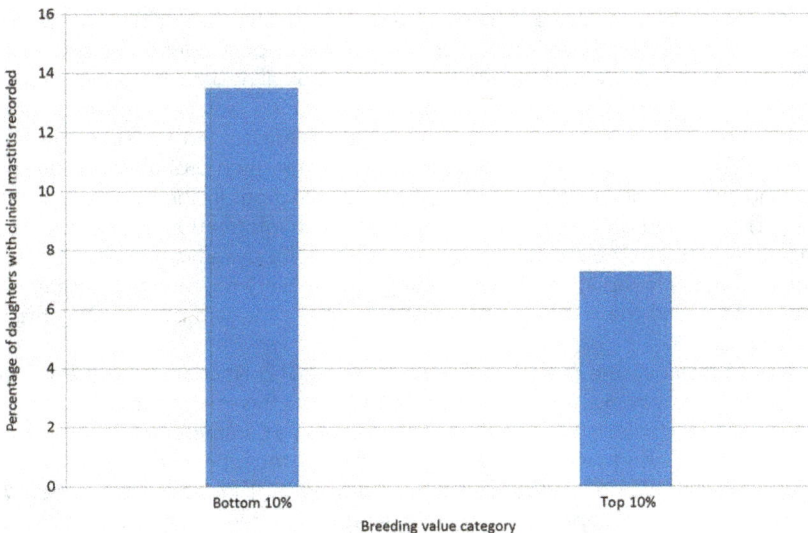

Figure 2 Percentage of recorded cases of clinical mastitis in around 2000 sires that are divergent for Australian national genetic evaluations of somatic cell count breeding values.

© Burleigh Dodds Science Publishing Limited, 2017. All rights reserved.

in test-day SCC, (3) number of days until SCC reaches an upper or lower threshold, (4) differences between observed SCC and SCC expected under healthy conditions, (5) areas under (parts of) the lactation curve of SCC, (6) standard deviation of log test-day SCC during lactation, (7) rolling SCC averages and (8) peaks in SCC.

Windig et al. (2010) examined which combination of alternative SCC traits can be used best to reduce both CM and SCM. Their conclusion was that a combination of five SCC traits (i.e. average SCC in early (5–150d) and late (>150d) lactation, suspicion of infection based on increased SCC, extent of increased SCC and presence of a peak pattern in SCC) gave a high accuracy of 0.91, when the aim was to reduce CM directly. Using a set of SCC traits can, therefore, partly overcome absence of direct observations on CM. In fact, direct observations on CM add little to the accuracy of an index if five SCC traits are used. To conclude, the urgency to set up a full infrastructure to collect data on CM is not essential to achieving acceptable rates of genetic improvement in mastitis resistance. In the Netherlands, an udder health index based on a combination of alternative SCC traits was introduced in 2010 (Eding and de Jong 2010). In Ireland, a similar index will be introduced in 2016.

In addition to SCC, other predictors of clinical mastitis could be used to increase the accuracy of breeding values. Examples include udder conformation (Lund et al. 1994), electrical conductivity from automated milking systems (Norberg 2005) and lactate dehydrogenase, which is a potential biomarker for mastitis (Friggens et al. 2007b). Norberg (2005) had reservations on practical aspects of data collation of electrical conductivity, but also illustrated that it could be used in breeding programmes. Electrical conductivity requires specialist machinery to evaluate, whereas SCC and lactate dehydrogenase have to be analysed by laboratories.

4.1.2 Recovery from mastitis

The studies described so far have investigated the transition probability of developing mastitis, but none of them have focused on the probability of recovering from an infection. Franzen et al. (2012) presented an alternative longitudinal approach in which evaluation of mastitis is performed based on changes in SCC during lactation; these changes are captured by modelling transition probabilities between assumed states of mastitis and non-mastitis. The method simultaneously models the transition probability of developing mastitis and the probability of recovering from an infection. In their approach, Franzen et al. (2012) modelled both aspects to capture as much information as possible from the SCC lactation pattern. The model did indeed capture the dynamic nature of the disease by modelling mastitis liability and by including the recovery process and repeated cases into the analysis, and the results point towards a significant gain by including the whole disease course.

One way to improve the model of Franzen et al. (2012) is to describe the fluctuating behaviour of SCC in milk in an online monitoring tool with a Bayesian approach to time series analysis (West and Harrison 1997). It is based on recursive parameter estimation, adaptive filtering for online (short-term) forecasting and offline retrospective analysis afterwards (backward smoothing) and change detection, followed by automatic intervention. André (2011) took the Bayesian approach to develop adaptive (self-learning) dynamic models for operational use in dairy farming.

This method will be able to describe fast (fluctuations around the level) and slow (changes in level itself) fluctuations of the SCC in milk, depending on the frequency of SCC sampling. Frequent sampling will, however, require a more advanced model. These

© Burleigh Dodds Science Publishing Limited, 2017. All rights reserved.

dynamic linear models can be used to describe individual SCC patterns and are able to detect process disturbances, such as (sub)clinical mastitis based on the SCC. It will help to detect a change in SCC level, and will simultaneously predict if a cow is capable of reducing SCC to her expected level after SCC increase, or if she needs treatment with antibiotics to combat the infection.

4.2 Lameness

While mastitis is the most common disease in dairy cattle, there are other diseases that should be considered as candidates for genetic improvement. After mastitis, the next most common welfare problem in dairy cattle is usually considered to be lameness, which is a major health and welfare issue for dairy cattle worldwide, and feet and leg issues are common reasons for culling in dairy cattle (Egger-Danner et al. 2015). Lameness results in increased veterinary costs, higher culling rates, reduced fertility and economic costs due to lower milk production and loss of body condition and live weight (Chawala et al. 2013). In Ireland, an overall incidence of 15% is assumed, of which 3% require veterinary intervention with associated higher costs (Berry and Amer 2005). Several studies have shown that lameness is heritable, with heritability estimates ranging from 0.07 to 0.10 reviewed by Kougioumtzis et al. (2014). Although the heritabilities are low, reduced lameness may be achieved by breeding for improved claw health. Clinical records on lameness are not always present, and selection to reduce lameness has historically focused on conformation of feet of legs, as this is routinely recorded by breed societies and many countries already calculate breeding values for these traits. However, the accuracy of breeding values for claw health or resistance to lameness increased when claw health data is included (Koenig et al. 2005).

A preliminary study in Ireland showed that there is a great benefit if direct information on lameness records is added to an index to predict lameness in addition to conformation traits. One way of accessing large amounts of clinical lameness data is to work with professional hoof trimmers who have expertise in diagnosis of claw disorders. An example has been developed by CRV (Arnhem, the Netherlands), who have set up a system, called DigiClaw (http://www.slideserve.com/mannix-sanford/implementation-of-a-claw-health-index-in-the-netherlands) together with the Animal Health Service in the Netherlands.

Another promising approach is to develop breeding values for different types of lameness, as there is evidence to suggest that heritabilities vary among claw diseases recorded by hoof trimmers (Buch et al. 2011; Ødegård et al. 2013). To do this on a level that is large enough for breeding value development (e.g. national) requires accurate and consistent data records. In fact, there has been a lot of effort recently to harmonise recording of claw disorder, for example the ICAR claw health atlas (Egger-Danner et al. 2014). Another option is to use locomotion scores that vary between normal gait and severely lame, which can be evaluated on all cows, but are usually only available once in lactation (Boelling and Pollott 1998).

4.3 Metabolic diseases

Metabolic disorders such as ketosis, displaced abomasum, milk fever and tetany are disturbances to one or more of the metabolic processes in dairy cattle. Intense selection for production has led to a reliance on body reserves to support early lactation. Consequently, the commencement of lactation and some of the remainder of lactation are often in negative energy balance. This leads to an imbalance in hormones and metabolites giving

© Burleigh Dodds Science Publishing Limited, 2017. All rights reserved.

rise to metabolic diseases (White 2015). Dysfunction or imbalance in metabolic processes leads to disease, so it is not surprising that genetic correlations between many dairy cow production diseases and milk production traits are mostly unfavourable (e.g. Uribe et al. 1995; Pryce et al. 1997; Zwald et al. 2004). Usable genetic variation in metabolic stability implies that breeding should be considered as a way to achieve improvements. Under-recording and difficulty in diagnosing sub-clinical cases are among the reasons why there is growing interest in using easily measurable predictors of metabolic diseases, either recorded 'on-farm' by using sensors and milk tests or recorded 'off-farm' using data collected from routine milk recording. Some countries have already initiated genetic evaluations of metabolic diseases (e.g. the Nordic countries and Austria/Germany) and these evaluations are based on clinical observations of disease.

4.4 Mid-infrared spectral data

One of the most promising ways of evaluating sub-clinical disease is the mid-infrared (MIR) analysis of milk samples. In addition to traditional traits (i.e. fat, protein, casein, lactose and urea contents), MIR analysis of milk has been used to predict other milk characteristics such as fatty acid composition, milk protein composition, milk coagulation properties, milk acidity, mineral composition and ketone bodies (De Marchi et al. 2014). For some of these traits such as ketone bodies, the accuracy of prediction is not high enough to use MIR predicted values as a reference value. However, the accuracy is sufficient for a rough screening to distinguish cows with high or low values. Hence, MIR may be an opportunity to massively increase the number of phenotypic records available for sub-clinical diseases, as MIR is used in standard milk analysis undertaken by milk recording organisations.

As with metabolic disease biomarkers, there is growing evidence that MIR can also be used to predict energy balance (McParland et al. 2014), which can be explained because catabolism of stored adipose reserves during body condition score change results in an increase in C18 fatty acid concentration in milk (Berry et al. 2013). Already there are several papers clearly showing that MIR can be used to predict the concentration of several fatty acids (Soyeurt et al. 2006; De Marchi et al. 2014) and might therefore mirror body condition score changes.

4.5 General disease resistance

Another appealing strategy is to select for more general immunity to diseases. Heringstad et al. (2007) showed that selection against mastitis leads to favourable correlated responses to selection in other diseases, such as ketosis and retained placenta, indicating the existence of a general robustness or reduced liability to disease. Furthermore, in a study by De la Paz (2008) it was reported that cows with both high antibody and cell-mediated immune response have a decreased risk of disease occurrence for several diseases, including mastitis, ketosis, metritis and retained placenta, compared with cows identified as low responders. The heritability of response to an immunity challenge is high enough to justify selection (Thompson-Crispi et al. 2012b). In fact, selection tools for immunity are available commercially. Semex (www.semex.com) sells semen from bulls identified as being high and low antibody and cell-mediated responders to an immune challenge. The high responders were found to have half the disease occurrence compared with low responders (Thompson-Crispi et al. 2012a).

© Burleigh Dodds Science Publishing Limited, 2017. All rights reserved.

5 New breeding objectives: dairy cows and climate change

Climate change is a growing international concern and it is well established that the release of greenhouse gases (GHG) is a contributing factor. At the recent Conference of Parties in Paris 2015 (www.cop21paris.org), many countries committed themselves to reduce their GHG emissions by 30% by year 2030 relative to 1990 levels. The global livestock sector, particularly ruminants, contributes approximately 18% of total anthropogenic GHG emissions (Steinfeld et al. 2006). Although mitigating the impact of ruminants on GHG emissions is not a welfare problem itself, the impacts of climate change on livestock poses a threat. Consequently, steps need to be taken to (1) reduce GHG emissions and (2) improve adaptability of cattle to climate change as a result of increased GHG emissions.

5.1 Greenhouse gas emissions

Enteric methane is produced as a by-product of anaerobic fermentation (methanogenesis) in the digestive tract by microorganisms called methanogens. Considering that milk production traits are a large part of the breeding goal, the GHG emitted per litre of milk are diluted in higher yielding cows. Consequently, it is worthwhile quantifying the impact of current selection on mitigation of GHG emissions. We estimate that under current rates of genetic gain in Australia, the amount of milk solids (fat plus protein) produced per cow will increase by 1.92 kg/year through selection and therefore we project that milk solids' yields will increase from 501 kg/cow/year to 520 kg/cow/year. If all cows in the current population (assumed to be 1.7 million) improve their yields by this amount then we estimate that approximately 60 000 fewer cows would be required to produce the same amount of milk solids. This is equivalent to a reduction of 397 383 t CO_2-eq. The second area of improvement is the reduction in emissions per unit of milk solids, that is, a dilution effect on emissions of having more productive cows of 33.62 gCO_2/kgMS/year, which is equivalent to 293 096 t CO_2-eq. After 10 years, the total annual impact of dairy selection practices on GHG emissions is projected to be 397 383 + 293 096 = 640 483 t/year, which is about 4.6% of the total current annual dairy emissions of 14 900 000 t/year. Selecting on other traits that improve the efficiency of farm systems, for example, milk yield, residual feed intake and longevity will also have a favourable effect on overall emissions (Wall et al. 2010).

Additional benefits could be achieved in reducing methane emissions if breeding values for methane emissions could be developed. However, building a sufficiently large dataset for genetic parameter estimation has been challenging, as phenotype data is scarce. New methods of detecting methane emissions are being developed, including gas sensors and radioactive tracers (SF_6), which will enable enough phenotypes to be collected to estimate genetic parameters. For example, using a portable air sampler and analyser unit to measure methane emissions on 3121 cows from 20 herds, Lassen and Løvendahl (2016) estimated that the heritability of methane emissions varied between 0.16 (s.e. 0.04) and 0.21 (s.e. 0.06) for various methane emission traits. Including methane emissions in the selection objective may further reduce greenhouse gas emissions at a small economic cost.

© Burleigh Dodds Science Publishing Limited, 2017. All rights reserved.

5.2 Heat tolerance

Animals have a comfort zone where body heat is effectively dissipated and the physiological state is maintained. When environmental parameters (e.g. temperature, humidity, radiation, solar and wind speed) go beyond this thermo-neutral zone (threshold), animals will start to experience heat load; if this becomes acute, heat stress will occur. Inability of animals to regulate body temperature under heat stress can result in loss of production, decrease in feed efficiency, suppression of immune system leading to increased susceptibility to diseases and decreased fertility (De Rensis and Scaramuzzi 2003). Differences in ability to cope with heat stress are influenced by several factors varying from animal characteristics (e.g. age, level of production) and physical properties (e.g. size, skin, coat) to environmental and herd management (e.g. feeding, housing, heat duration and abatement techniques). The ability to cope with heat stress varies among breeds: Holsteins appear to exhibit greater reductions in milk yield in hotter climates than Jerseys or cross-breeds (Bryant et al. 2007). In fact, reductions in yield start when temperatures exceed 21 and 25°C at 75% humidity for Holsteins and Jerseys, respectively (Bryant et al. 2007). Nguyen et al. (2016) used a similar approach to calculate genetic parameters for heat tolerance and estimated that the heritability of heat tolerance was around 0.11.

6 Genomic selection, inbreeding and gene editing

6.1 Genomic selection

There are very few examples where the price of a commodity has dropped as dramatically as genotyping. In 2001, the first human being was sequenced at a price of US$3 billion (Venter et al. 2001). Since then, the $1000 genome has become a bit of a catchphrase and is now a reality. One of the biggest successes in agricultural science in recent years has been to leverage off investment made in medical genetics and applying it to make smart new ways to select the best individuals for the next generation. Sequencing of bulls (key ancestors of dairy and beef breeds) is now happening around the world, with the 1000 Bulls Genome Project comfortably exceeding that target and offering new insights into genetic architecture in addition to increasing the accuracy of genomic prediction of many other traits, by providing more informative genetic markers to use in genomic selection (Daetwyler et al. 2014).

Genomic selection refers to selection decisions based on genomic breeding values (Meuwissen et al. 2001). The way it works is that a genomic reference population is assembled, typically genotyped bulls with large numbers of progeny. A genomic prediction equation is then calculated from the reference population by looking for associations between phenotypes and dense genetic markers that are approximately equally spaced across the entire genome. In cattle a variety of marker panels are used, varying from a few thousand genetic markers per panel to hundreds of thousands. The genomic breeding values are calculated as the sum of each of the genotypes multiplied by its respective effect on a trait, thereby potentially capturing most of the genes that cause differences among animals in the traits of interest. This prediction equation can then be applied to individuals that are genotyped, but have no phenotypes. Therefore, the genetic merit of an individual can be calculated as early as birth and, therefore, selection decisions can be made earlier in life than traditional progeny-test approaches.

© Burleigh Dodds Science Publishing Limited, 2017. All rights reserved.

Genomic selection is now used routinely in many countries for genetic evaluation of traits that already have an estimated breeding value derived from a combination of pedigree and phenotype information (Spelman et al. 2013). The advantage of genomic selection for these traits is that the rate of genetic gain is accelerated by 40–50% (Spelman et al. 2013).

To date, there have been limited attempts to calculate genomic predictions for health traits; countries that include genomic information in their genetic prediction of health disorders include Canada, France and Scandinavia. Other countries, such as the United States are at the time of writing in the research phase, but have found that genomic information improves the accuracy of prediction.

Genomic reference populations may assist with difficult-to-measure traits, such as health, as efforts to record and evaluate these traits can happen in a small reference population and the benefits used by the entire population, that is, prediction equations are based on cows in the reference population that have phenotypes on a range of traits, possibly also including health traits. For cheap and easy to measure phenotypes, reasonable reliabilities can be achieved using reference populations comprising genotyped bulls with progeny groups. For traits that are expensive to measure, or where data is sparse, the best option is to obtain phenotypes on genotyped cows (Chesnais et al. 2016).

In some circumstances, adding females to the reference population can be advantageous for traits that are measured on a large (national) scale, for example, production, longevity and fertility. It appears that the biggest gains are made when the genotyped females comprise a high percentage of the overall reference population and where the cows are selected based on very-high-quality phenotypes.

An example of adding females to the reference population is the Australian Ginfo population of around 25 000 Holsteins, Jerseys and cross-bred cows from 100 herds that were selected based on objective criteria around recording quality. By adding the Ginfo cows to the reference population, the size of the total Holstein and Jersey reference populations increased by 44% and 38%, respectively. The reliability of the BPI increased by 5.8% in Holstein genotyped animals; the reliability of fertility breeding values improved by 4/5% and overall type improved by 7.1% (ADHIS, 2016; unpublished results). These are substantial improvements and pave the way for extending the number of traits evaluated, as the future Ginfo population is expected to be a rich resource in phenotypes for 'new' traits.

The next step in future genomic evaluations is to use information from biological priors to further improve the accuracy of selection. Genome-wide association studies (GWAS) are often used to identify regions of the genome that have a specific impact on a trait of economic importance. Sometimes these can be linked to candidate genes, which helps our understanding of the genetic control of complex traits and if large enough, can also be selected on directly. GWAS have already been used to identify parts of the genome with large effect on fertility and some promising candidate genes have been identified (Sahana et al. 2010; Pimentel et al. 2011). Whole-genome sequencing is likely to improve the resolution and detection of causative mutations of these candidate SNPs. While new mutations are likely to be identified, the challenge will be to apply this knowledge. One option is to use this as prior information in genomic selection methods. There are examples where this has been tested and increases in accuracy of prediction achieved. For example, Khansefid et al. (2014) found that placing more emphasis on SNP associated with residual feed intake in beef cattle increased the accuracy of genomic prediction in dairy cattle.

© Burleigh Dodds Science Publishing Limited, 2017. All rights reserved.

6.2 Inbreeding

Recording of ancestry has been the cornerstone of genetic improvement programmes in livestock, with breeding decisions historically being made on breeding values estimated using a combination of pedigree and phenotype data. Extensive use of artificial insemination and very similar worldwide selection objectives has led to intense selection of superior males to become sires of the next generation. Therefore, it is almost impossible to find dairy animals without genetic ties to certain key ancestor bulls. This has led to an increase in inbreeding reported in most dairy populations (Miglior et al. 1995; Wiggans et al. 1995; VanRaden et al. 2011).

Another use of genomic data is to monitor inbreeding in a population by quantifying genomic relationships between animals (Pryce et al. 2012). Inbreeding arises in individuals that have parents that share a common ancestor, which is known as identity by descent. Pedigree is commonly used to assess the inbreeding of the individual itself and its relationship to others in the population.

As inbreeding increases, regions of the genome become more homozygous and also increases the risk of homozygous lethal recessives. There are examples of genetic diseases that are lethal recessives, such as CVM, BLAD and DUMPS in Holsteins. Most of these diseases are the result of reasonably recent (rare) mutations. For example complex vertebral malformation, or CVM, can be traced to two former elite Holstein sires; as a result of their widespread use, the sires appeared on both sides of the pedigree of affected calves (Agerholm et al. 2001). Conditions such as these diseases highlight the importance of managing rates of inbreeding, which arises as a result of the co-occurrence of common ancestor(s) in maternal and paternal pedigrees.

An example of identity by descent is shown in Fig. 3.

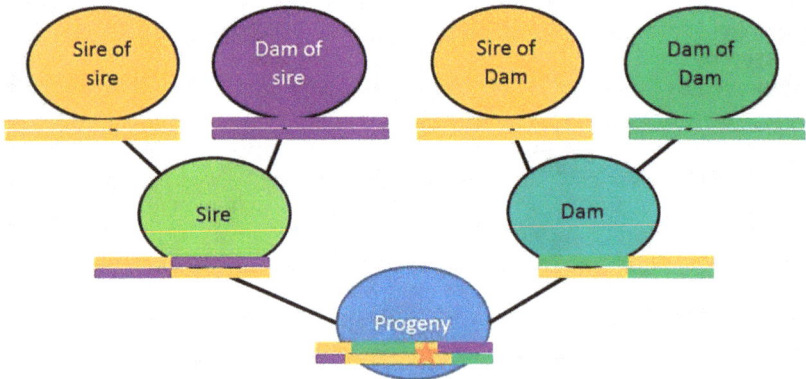

Figure 3 A case study in the use of genomics to identify homozygosity and identity by descent. The individual (progeny) that results from the mating of the sire and dam is inbred, as the grand-sire on either side of the pedigree is the same animal (shown in yellow). If the genomic regions from the grand-sire are tracked from generation to generation, the segments that are proportion of homozygous can be quantified. If the grand-sire is a carrier of a lethal recessive, the progeny will be affected if it inherits both copies of the recessive.

© Burleigh Dodds Science Publishing Limited, 2017. All rights reserved.

6.3 Gene editing

An opportunity to use genetics to improve animal welfare is through new technologies, such as gene editing. Gene-editing techniques can be used to precisely alter the genome through inserting, editing or DNA sequencing (Hsu et al. 2014). Gene editing makes it possible to change or disable a single gene without changing the 'meaning' of the rest of the genome. This means that undesired effects such as accidentally turning off a useful gene are less likely than with previous genome modification techniques and the desired gene can be introduced rapidly into a population. However, there are regulation issues associated with applying gene editing to livestock that need to be dealt with before marketable applications are made. Provided the regulation issues can be overcome, there are many applications that could have major implications for animal welfare. Here, we will provide three examples that impact on animal welfare in quite different ways: the first example is introgression of the polled gene (hornless cattle), the second example is the cholesterol deficiency mutation associated with juvenile mortality and the third is the Slick gene for improved heat tolerance.

Many of the traits that we deal with in animal breeding are polygenic in their genetic control. This means that the genetic architecture is often complex, involving many loci. Although it may be obvious that quantitative traits, such as height or weight are controlled by multiple loci, many diseases are also complex in their genetic architecture. Although we have given examples of single genes that can be edited, multiple sites can be edited simultaneously (Hickey et al. 2016).

6.3.1 Polled gene

Most dairy heifers are disbudded or dehorned at an early age. The procedure is generally done using heat cauterisation, often without the use of anaesthetic and is therefore considered an animal welfare issue. The gene for polled is a single dominant gene. Therefore, mating a homozygous polled bull (PP) to a herd of non-polled cows (hh) will result in all the offspring being polled (Ph). If the bull is heterozygous (Ph) and the cows are horned (hh), half the offspring will be polled (Ph). Two mutations that prevent development of horns in certain breeds of cattle have been mapped on the bovine genome (Medugorac et al. 2012). Genetic dehorning of cattle would certainly be preferable, however, introgression of the polled gene through conventional selection would lead to a trade-off in genetic merit, as carriers of the polled mutation are generally inferior in genetic merit, recovery of genetic merit could take several generations. Gene editing is an obvious solution and a method to do this using the gene-editing technique called transcription activator-like effector nucleases has already been used to generate live polled calves (Fahrenkrug and Carlson 2014). The advantage with this method is that no additional genetic material is transferred between breeds.

6.3.2 Cholesterol deficiency

In 2015, a new defect was discovered in Holsteins by Kipp et al. (2015) at VIT in Germany that causes young calves to die if homozygous. They concluded that heterozygous animals have reduced cholesterol, but homozygotes have no cholesterol and survive only a few months. The defective haplotype traces back to one sire born in 1991 (https://www.cdcb.us/reference/changes/HCD_inheritance.pdf). This indicates that animals that have this sire in their pedigree more than two times might be carriers of this defect (see Fig 3). Through

© Burleigh Dodds Science Publishing Limited, 2017. All rights reserved.

the availability of genomic information, this defect is known, but it also provides a basis for genome-assisted approaches to avoiding inadvertent carrier matings.

6.3.3 Slick gene

As mentioned earlier in this chapter, heat stress is a big problem for the lactating cow. One approach for improving resistance to heat stress in dairy breeds is to introduce thermotolerance genes from other breeds. One such gene is called the slick gene. The slick is described as a very short, sleek hair coat mostly observed in tropical *Bos taurus* cattle of Criollo descent (Huson et al. 2014). Slick cattle are better able to regulate body temperature during heat stress than cows with normal hair (Dikmen et al. 2014).

7 Summary

Genetic selection has led to phenomenal rates of genetic gain in milk production traits. However, from the mid-1990s, it was recognised that narrow breeding goals, focused on production traits, had negative consequences for fitness traits, impacting on animal welfare. The most obvious was the deterioration in female fertility, which has been observed worldwide. Since then, breeding goals have been extended and realised selection responses for traits such as fertility show that genetic selection can improve even low heritability traits. More recently, there has been a surge of interest in selecting for health traits and traits associated with resources, such as feed efficiency and methane emissions, even though the collection of these phenotypes may be expensive and only possible in small populations. In fact, all these traits are ideal candidates for genomic selection where the basic requirement is a population that has genotypes and phenotypes (either on the individual or on its progeny). Genotypes and associated phenotypes are used to create a reference population to calculate genomic prediction equations that can be applied to cows or bulls that are genotyped but do not have phenotypes. While genomic selection has been implemented for many traits (production, fertility, cell count, longevity, etc.), there are still obstacles in applying it to several new traits, associated with the heritability of the trait, the number of animals in the reference population and the cost of phenotyping. The opportunities to apply genomic selection to traits related to welfare are abundant and include common health disorders, such as mastitis resistance, lameness and ketosis. One option is to customise future breeding goals to include appropriate traits for the prevailing management and environmental conditions. For example, selection for heat tolerance is of great importance in emerging (and many existing) dairy regions. In conclusion, genomic selection offers exciting prospects for genetically improving scarce and expensive to measure traits, such as those associated with dairy welfare and resource usage.

8 Where to look for further information

Many new and relevant journal articles are published by the *Journal of Dairy Science* (JDS) including a recent collection of papers on balanced breeding: http://www.journalofdairyscience.org/content/balancedbreeding.

Also Interbull has annual meetings and freely provides the latest finding and discussions: http://www.interbull.org.

© Burleigh Dodds Science Publishing Limited, 2017. All rights reserved.

9 Acknowledgements

Jennie Pryce acknowledges the Gardiner Foundation (Melbourne, Australia) for financial support through the ImProving Herds project. We acknowledge Professor Ben Hayes (University of Queensland) and Dr Matthew Bell (University of Nottingham) for their contribution to the sections on adaptability to climate change. We also thank Dr Mary Abdelsayed (Holstein Australia, Melbourne) for providing us with unpublished data contributing to Fig. 2.

10 References

Agerholm, J. S., Bendixen, C., Andersen, O. and Arnbjerg, J. (2001), Complex vertebral malformation in Holstein calves. *Journal of Veterinary Diagnostic Investigation* 13, 283–9.

André, G. (2011), *Adaptive Models for Operational Use in Dairy Farming – Increasing Economic Results Utilising Individual Variation in Response.* Wageningen University, Wageningen, the Netherlands.

Banos, G., Coffey, M. and Brotherstone, S. (2005), Modeling daily energy balance of dairy cows in the first three lactations. *Journal of Dairy Science* 88, 2226–37.

Beilharz, R., Luxford, B. and Wilkinson, J. (1993), Quantitative genetics and evolution: is our understanding of genetics sufficient to explain evolution? *Journal of Animal Breeding and Genetics* 110, 161–70.

Berry, D., McParland, S., Bastin, C., Wall, E., Gengler, N. and Soyeurt, H. (2013), Phenotyping of robustness and milk quality. *Advances in Animal Biosciences* 4, 600.

Berry, D., Wall, E. and Pryce, J. (2014), Genetics and genomics of reproductive performance in dairy and beef cattle. *Animal* 8, 105–21.

Berry, D. P. and Amer, P. R. (2005), *Derivation of a Health Sub-Index for the Economic Breeding Index in Ireland.* Technical report to the Irish Cattle Breeding Federation, Cork, Ireland.

Bewley, J., Peacock, A., Lewis, O., Boyce, R., Roberts, D., Coffey, M., Kenyon, S. and Schutz, M. (2008), Potential for estimation of body condition scores in dairy cattle from digital images. *Journal of Dairy Science* 91, 3439–53.

Boelling, D. and Pollott, G. (1998), Locomotion, lameness, hoof and leg traits in cattle II: Genetic relationships and breeding values. *Livestock Production Science* 54, 205–15.

Boichard, D. and Brochard, M. (2012), New phenotypes for new breeding goals in dairy cattle. *Animal* 6, 544–50.

Bryant, J., López-Villalobos, N., Pryce, J., Holmes, C. and Johnson, D. (2007), Quantifying the effect of thermal environment on production traits in three breeds of dairy cattle in New Zealand. *New Zealand Journal of Agricultural Research* 50, 327–38.

Buch, L. H., Sørensen, A. C., Lassen, J., Berg, P., Eriksson, J.-Å., Jakobsen, J. and Sørensen, M. K. (2011), Hygiene-related and feed-related hoof diseases show different patterns of genetic correlations to clinical mastitis and female fertility. *Journal of Dairy Science* 94, 1540–51.

Chawala, A. R., Lopez-Villalobos, N., Margerison, J. K. and Spelman, R. J. (2013), Genetic and crossbreeding parameters for incidence of recorded clinical lameness in New Zealand dairy cattle. *New Zealand Veterinary Journal* 61, 281–5.

Chesnais, J., Cooper, T., Wiggans, G., Sargolzaei, M., Pryce, J. and Miglior, F. (2016), Using genomics to enhance selection of novel traits in North American dairy cattle. *Journal of Dairy Science* 99, 2413–27.

Daetwyler, H. D., Capitan, A., Pausch, H., Stothard, P., Van Binsbergen, R., Brøndum, R. F., Liao, X., Djari, A., Rodriguez, S. C. and Grohs, C. (2014), Whole-genome sequencing of 234 bulls facilitates mapping of monogenic and complex traits in cattle. *Nature Genetics* 46, 858–65.

© Burleigh Dodds Science Publishing Limited, 2017. All rights reserved.

de Haas, Y., Barkema, H. W., Schukken, Y. H. and Veerkamp, R. F. (2003), Genetic associations for pathogen-specific clinical mastitis and patterns of peaks in somatic cell count. *Animal Science* 77, 187–95.

De la Paz, J. M. (2008), *Using Antibody and Cell-Mediated Immune Response to Test Antigens in Periparturient Dairy Cows as a Measure of Disease Resistance.* University of Florida.

De Marchi, M., Toffanin, V., Cassandro, M. and Penasa, M. (2014), Invited review: mid-infrared spectroscopy as phenotyping tool for milk traits. *Journal of Dairy Science* 97, 1171–86.

De Rensis, F. and Scaramuzzi, R. J. (2003), Heat stress and seasonal effects on reproduction in the dairy cow – a review. *Theriogenology* 60, 1139–51.

De Vries, M. and Veerkamp, R. (2000), Energy balance of dairy cattle in relation to milk production variables and fertility. *Journal of Dairy Science* 83, 62–9.

Detilleux, J. C., Jacquinet, E., Harvengt, A. and Leroy, P. L. (1997), Genetic selection for resistance to mastitis. *Annales De Medecine Veterinaire* 141, 199–&.

Dikmen, S., Khan, F. A., Huson, H. J., Sonstegard, T. S., Moss, J. I., Dahl, G. E. and Hansen, P. J. (2014), The SLICK hair locus derived from Senepol cattle confers thermotolerance to intensively managed lactating Holstein cows. *Journal of Dairy Science* 97, 5508–20.

Eding, E. and de Jong, G. (2010), Predicting mastitis resistance breeding values from somatic cell count indicator traits. In *World Conference of Genetics Applied to Livestock Production (WCGALP)*, Leipzig, Germany.

Egger-Danner, C., Cole, J., Pryce, J., Gengler, N., Heringstad, B., Bradley, A. and Stock, K. (2015), Invited review: overview of new traits and phenotyping strategies in dairy cattle with a focus on functional traits. *Animal* 9, 191–207.

Egger-Danner, C., Nielsen, P., Fiedler, A., Müller, K., Fjeldaas, T., Döpfer, D., Daniel, V., Bergsten, C., Cramer, G. and Christen, A. (2014), ICAR Claw Health Atlas. *ICAR Technical Series.* ICAR Working Group on Functional Traits (ICAR WGFT) and International Claw Health Experts, pp. 6–7.

Emanuelson, U. (1988), Recording of production diseases in cattle and possibilities for genetic improvements: a review. *Livestock Production Science* 20, 89–106.

Fahrenkrug, S. C. and Carlson, D. F. (2014), *Hornless Livestock.* Google Patents.

Franzen, J., Thorburn, D., Urioste, J. I. and Strandberg, E. (2012), Genetic evaluation of mastitis liability and recovery through longitudinal analysis of transition probabilities. *Genetics Selection Evolution* 44, 10.

Friggens, N., Berg, P., Theilgaard, P., Korsgaard, I. R., Ingvartsen, K. L., Løvendahl, P. and Jensen, J. (2007a), Breed and parity effects on energy balance profiles through lactation: evidence of genetically driven body energy change. *Journal of Dairy Science* 90, 5291–305.

Friggens, N., Chagunda, M., Bjerring, M., Ridder, C., Hojsgaard, S. and Larsen, T. (2007b), Estimating degree of mastitis from time-series measurements in milk: a test of a model based on lactate dehydrogenase measurements. *Journal of Dairy Science* 90, 5415–27.

Gaddis, K. P., Cole, J., Clay, J. and Maltecca, C. (2014), Genomic selection for producer-recorded health event data in US dairy cattle. *Journal of Dairy Science* 97, 3190–9.

Green, M. J., Green, L. E., Schukken, Y. H., Bradley, A. J., Peeler, E. J., Barkema, H. W., de Haas, Y., Collis, V. J. and Medley, G. F. (2004), Somatic cell count distributions during lactation predict clinical mastitis. *Journal of dairy science* 87, 1256–64.

Hazel, L. N. (1943), The genetic basis for constructing selection indexes. *Genetics* 28, 476–90.

Heringstad, B., Gianola, D., Chang, Y., Ødegård, J. and Klemetsdal, G. (2006), Genetic associations between clinical mastitis and somatic cell score in early first-lactation cows. *Journal of Dairy Science* 89, 2236–44.

Heringstad, B., Klemetsdal, G. and Steine, T. (2007), Selection responses for disease resistance in two selection experiments with Norwegian red cows. *Journal of dairy science* 90, 2419–26.

Heringstad, B. and Østerås, O. (2013), More than 30 years of health recording in Norway. *ICAR Technical Series no. 17*, 39.

Heuven, H. C. M. (1987), *Diagnostic and Genetic Analysis of Mastitis Field Data.* University of Wisconsin, Madison, US.

© Burleigh Dodds Science Publishing Limited, 2017. All rights reserved.

Hickey, J., Bruce, C., Whitelaw, A. and Gorjanc, G. (2016), Promotion of alleles by genome editing in livestock breeding programmes. *Journal of Animal Breeding and Genetics* 133, 83–4.

Hsu, P. D., Lander, E. S. and Zhang, F. (2014), Development and applications of CRISPR-Cas9 for genome engineering. *Cell* 157, 1262–78.

Huson, H. J., Kim, E.-S., Godfrey, R. W., Olson, T. A., McClure, M. C., Chase, C. C., Rizzi, R., O'Brien, A. M., Van Tassell, C. P. and Garcia, J. F. (2014), Genome-wide association study and ancestral origins of the slick-hair coat in tropically adapted cattle. *Frontiers in genetics*, 5, 101.

Kelton, D. F., Lissemore, K. D. and Martin, R. E. (1998), Recommendations for recording and calculating the incidence of selected clinical diseases of dairy cattle. *Journal of Dairy Science* 81, 2502–9.

Khansefid, M., Pryce, J., Bolormaa, S., Miller, S., Wang, Z., Li, C. and Goddard, M. (2014), Estimation of genomic breeding values for residual feed intake in a multibreed cattle population. *Journal of Animal Science* 92, 3270–83.

Kipp, S., Segelke, D., Schierenbeck, S., Reinhardt, F., Reents, R., Wurmser, C., Pausch, H., Fries, R., Thaller, G., Tetens, J., Pott, J., Piechotta, M. and Grünberg, W. (2015), *A New Holstein Haplotype Affecting Calf Survival*. Joint Animal Meeting, Orlando, Florida, US.

Knight, C. H., D. E. B. and A. S. (1999), Metabolic loads to be expected from different genotypes under different systems. *British Society of Animal Science Occasional Publication* 24, 27–36.

Koenig, S., Sharifi, A., Wentrot, H., Landmann, D., Eise, M. and Simianer, H. (2005), Genetic parameters of claw and foot disorders estimated with logistic models. *Journal of Dairy Science* 88, 3316–25.

Kougioumtzis, A., Valergakis, G. E., Oikonomou, G., Arsenos, G. and Banos, G. (2014), Profile and genetic parameters of dairy cattle locomotion score and lameness across lactation. *Animal* 8, 20–7.

Lassen, J. and Løvendahl, P. (2016), Heritability estimates for enteric methane emissions from Holstein cattle measured using noninvasive methods. *Journal of Dairy Science* 99, 1959–67.

Lund, T., Miglior, F., Dekkers, J. and Burnside, E. (1994), Genetic relationships between clinical mastitis, somatic cell count, and udder conformation in Danish Holsteins. *Livestock Production Science* 39, 243–51.

Lush, J. L. (1960), Improving dairy cattle by breeding. I. Current status and outlook. *Journal of Dairy Science* 43, 702–6.

Martin-Collado, D., Byrne, T., Amer, P., Santos, B., Axford, M. and Pryce, J. (2015), Analyzing the heterogeneity of farmers' preferences for improvements in dairy cow traits using farmer typologies. *Journal of Dairy Science* 98, 4148–61.

McParland, S., Lewis, E., Kennedy, E., Moore, S., McCarthy, B., O'Donovan, M., Butler, S., Pryce, J. and Berry, D. (2014), Mid-infrared spectrometry of milk as a predictor of energy intake and efficiency in lactating dairy cows. *Journal of Dairy Science* 97, 5863–71.

Medugorac, I., Seichter, D., Graf, A., Russ, I., Blum, H., Göpel, K. H., Rothammer, S., Förster, M. and Krebs, S. (2012), Bovine polledness–an autosomal dominant trait with allelic heterogeneity. *PLoS ONE* 7, e39477.

Meuwissen, T. H., Hayes, B. J. and Goddard, M. E. (2001), Prediction of total genetic value using genome-wide dense marker maps. *Genetics* 157, 1819–29.

Miglior, F., Burnside, E. B. and Kennedy, B. W. (1995), Production traits of Holstein cattle: estimation of nonadditive genetic variance components and inbreeding depression. *Journal of Dairy Science* 78, 1174–80.

Mrode, R. and Swanson, G. (1996), Genetic and statistical properties of somatic cell count and its suitability as an indirect means of reducing the incidence of mastitis in dairy cattle. *Animal Breeding Abstracts (United Kingdom)*.

Nguyen, T. T., Bowman, P. J., Haile-Mariam, M., Pryce, J. E. and Hayes, B. J. (2016), Genomic selection for tolerance to heat stress in Australian dairy cattle. *Journal of Dairy Science* 99, 2849–62.

Nielsen, H.-M., Christensen, L. G. and Groen, A. (2005), Derivation of sustainable breeding goals for dairy cattle using selection index theory. *Journal of Dairy Science* 88, 1882–90.

© Burleigh Dodds Science Publishing Limited, 2017. All rights reserved.

Norberg, E. (2005), Electrical conductivity of milk as a phenotypic and genetic indicator of bovine mastitis: a review. *Livestock Production Science* 96, 129–39.

Ødegård, C., Svendsen, M. and Heringstad, B. (2013), Genetic analyses of claw health in Norwegian Red cows. *Journal of Dairy Science* 96, 7274–83.

Pimentel, E., Bauersachs, S., Tietze, M., Simianer, H., Tetens, J., Thaller, G., Reinhardt, F., Wolf, E. and König, S. (2011), Exploration of relationships between production and fertility traits in dairy cattle via association studies of SNPs within candidate genes derived by expression profiling. *Animal Genetics* 42, 251–62.

Pryce, J., Hayes, B. and Goddard, M. (2012), Novel strategies to minimize progeny inbreeding while maximizing genetic gain using genomic information. *Journal of Dairy Science* 95, 377–88.

Pryce, J. and Veerkamp, R. (2001), The incorporation of fertility indices in genetic improvement programmes. *British Society of Animal Science* 26, 237–49.

Pryce, J., Veerkamp, R., Thompson, R., Hill, W. and Simm, G. (1997), Genetic aspects of common health disorders and measures of fertility in Holstein Friesian dairy cattle. *Animal Science* 65, 353–60.

Pryce, J. E., Coffey, M. P. and Simm, G. (2001), The relationship between body condition score and reproductive performance. *Journal of Dairy Science* 84, 1508–15.

Pryce, J. E., Woolaston, R., Berry, D. P., Wall, E., Winters, M., Butler, R. and Shaffer, M. (2014), World trends in dairy cow fertility. *Proceedings 10th World Congress of Genetics Applied to Livestock Production.* https://asas.org/docs/default-source/wcgalp-proceedings-oral/154_paper_10356_manuscript_1630_0.pdf?sfvrsn=2.

Rauw, W., Kanis, E., Noordhuizen-Stassen, E. and Grommers, F. (1998), Undesirable side effects of selection for high production efficiency in farm animals: a review. *Livestock Production Science* 56, 15–33.

Reents, R., Dekkers, J. C. M. and Schaeffer, L. R. (1995), Genetic evaluation for somatic cell score with a test day model for multiple lactations. *Journal of Dairy Science* 78, 2858–70.

Roche, J. R., Friggens, N. C., Kay, J. K., Fisher, M. W., Stafford, K. J. and Berry, D. P. (2009), Invited review: body condition score and its association with dairy cow productivity, health, and welfare. *Journal of Dairy Science* 92, 5769–801.

Sahana, G., Guldbrandtsen, B., Bendixen, C. and Lund, M. (2010), Genome-wide association mapping for female fertility traits in Danish and Swedish Holstein cattle. *Animal Genetics* 41, 579–88.

Schepers, A. J., Lam, T., Schukken, Y. H., Wilmink, J. B. M. and Hanekamp, W. J. A. (1997), Estimation of variance components for somatic cell counts to determine thresholds for uninfected quarters. *Journal of Dairy Science* 80, 1833–40.

Soyeurt, H., Dardenne, P., Dehareng, F., Lognay, G., Veselko, D., Marlier, M., Bertozzi, C., Mayeres, P. and Gengler, N. (2006), Estimating fatty acid content in cow milk using mid-infrared spectrometry. *Journal of Dairy Science* 89, 3690–5.

Spelman, R. J., Hayes, B. J. and Berry, D. P. (2013), Use of molecular technologies for the advancement of animal breeding: genomic selection in dairy cattle populations in Australia, Ireland and New Zealand. *Animal Production Science* 53, 869–75.

Steinfeld, H., Gerber, P., Wassenaar, T., Castel, V., Rosales, M. and de Haan, C. (2006), *Livestock's Long Shadow: Environmental Issues and Options.* Food and Agriculture Organization of the United Nations, Rome, Italy.

Thompson-Crispi, K., Hine, B., Quinton, M., Miglior, F. and Mallard, B. (2012a), Short communication: association of disease incidence and adaptive immune response in Holstein dairy cows. *Journal of Dairy Science* 95, 3888–93.

Thompson-Crispi, K. A., Sewalem, A., Miglior, F. and Mallard, B. A. (2012b), Genetic parameters of adaptive immune response traits in Canadian Holsteins. *Journal of Dairy Science* 95, 401–9.

Thorup, V. M., Edwards, D. and Friggens, N. C. (2012), On-farm estimation of energy balance in dairy cows using only frequent body weight measurements and body condition score. *Journal of Dairy Science* 95, 1784–93.

© Burleigh Dodds Science Publishing Limited, 2017. All rights reserved.

Uribe, H., Kennedy, B., Martin, S. and Kelton, D. (1995), Genetic parameters for common health disorders of Holstein cows. *Journal of Dairy Science* 78, 421–30.

VanRaden, P. (2004), *Invited review*: selection on net merit to improve lifetime profit. *Journal of Dairy Science* 87, 3125–31.

VanRaden, P., Sanders, A., Tooker, M., Miller, R., Norman, H., Kuhn, M. and Wiggans, G. (2004), Development of a national genetic evaluation for cow fertility. *Journal of Dairy Science* 87, 2285–92.

VanRaden, P. M., Olson, K. M., Wiggans, G. R., Cole, J. B. and Tooker, M. E. (2011), Genomic inbreeding and relationships among Holsteins, Jerseys, and Brown Swiss. *Journal of Dairy Science* 94, 5673–82.

Veerkamp, R. (1998), Selection for economic efficiency of dairy cattle using information on live weight and feed intake: a review. *Journal of Dairy Science* 81, 1109–19.

Veerkamp, R., Beerda, B. and Van der Lende, T. (2003), Effects of genetic selection for milk yield on energy balance, levels of hormones, and metabolites in lactating cattle, and possible links to reduced fertility. *Livestock Production Science* 83, 257–75.

Venter, J. C., Adams, M. D., Myers, E. W., Li, P. W., Mural, R. J., Sutton, G. G., Smith, H. O., Yandell, M., Evans, C. A. and Holt, R. A. (2001), The sequence of the human genome. *Science* 291, 1304–51.

Wall, E., Simm, G. and Moran, D. (2010), Developing breeding schemes to assist mitigation of greenhouse gas emissions. *Animal* 4, 366–76.

West, M. and Harrison, J. (1997), *Bayesian Forecasting and Dynamic Models*. Springer-Verlag, New York, US.

White, H. M. (2015), The role of TCA cycle anaplerosis in ketosis and fatty liver in periparturient dairy cows. *Animals* 5, 793–802.

Wiggans, G. R., VanRaden, P. M. and Zuurbier, J. (1995), Calculation and use of inbreeding coefficients for genetic evaluation of United States dairy cattle. *Journal of Dairy Science* 78, 1584–90.

Windig, J. J., Ouweltjes, W., ten Napel, J., de Jong, G., Veerkamp, R. F. and De Haas, Y. (2010), Combining somatic cell count traits for optimal selection against mastitis. *Journal of Dairy Science* 93, 1690–701.

Zwald, N., Weigel, K., Chang, Y., Welper, R. and Clay, J. (2004), Genetic selection for health traits using producer-recorded data. I. Incidence rates, heritability estimates, and sire breeding values. *Journal of Dairy Science* 87, 4287–94.

© Burleigh Dodds Science Publishing Limited, 2017. All rights reserved.

Chapter 3

Improving smallholder dairy farming in tropical Asia

John Moran, Profitable Dairy Systems, Australia

1 Introduction

Globally, agriculture provides a livelihood for more people than any other industry (primary or secondary), while dairy farming is one of the major agricultural activities. The Food and Agriculture Organization (FAO) estimated that the world's milk production in 2012 stood at 754 billion tonnes. Hemme and Otto (2010) estimated that 12–14% of the world's population (or a total of 750–900 million people) live on dairy farms or are within dairy farming households. Livestock provide over half the value of global agricultural output and one third of the value of agricultural output in developing countries. Milk is nature's most complete food and dairy farming represents one of the fastest returns for livestock keepers in the developing world.

The Asia-Pacific region has seen the world's highest growth in demand for milk and dairy products over the last 30 years. Even though Asia has increased its milk output (as a percentage of global production) from 15% in 1981 to 37% in 2011, it still accounts for over 40% of the world's total dairy imports. The consumption of milk and dairy products in Asia has doubled over the last 30 years, now contributing to more than 60% of the total increase in global consumption.

Many of these countries now have school milk programmes to encourage young children to drink more milk and thus improve their health through increased consumption of the energy, protein and minerals (particularly calcium and phosphorus) contained in it.

http://dx.doi.org/10.19103/AS.2015.0005.37
© Burleigh Dodds Science Publishing Limited, 2017. All rights reserved.

In future years, as these children grow and have their own families, milk consumption will increase at an even faster rate. In the future, per capita milk consumption in South East Asia is expected to nearly double – from the current 10–12 kg/hd/yr to 19–20 kg/hd/yr by the year 2020 (Delgardo et al. 2003). This 3% per annum growth will lead to a total milk consumption of 12 million tonnes/yr by 2020, which Delgardo et al. (2003) predict will require a net import of 9 million tonnes of milk/yr. This will be a significant increase from the 4.7 million tonnes of milk/yr imported in 2000. In summary, by 2020, South East Asia will be producing only 25% of its total milk requirements.

Such growing demands have arisen due to a combination of the following factors:

- increasing per capita incomes;
- the emergence of affluent middle-class people in many low- to middle-income countries;
- westernisation trends which increase the demand for protein foods and value-added dairy products;
- increasing urbanisation; and
- expansion of modern retail outlets (with refrigeration cabinets) throughout Asia.

In other words, higher incomes and increasing urbanisation have combined with economic reforms and market liberalisation policies to heighten the import dependency of many countries in this region. Asia has then become increasingly dependent on the highly competitive, but increasingly volatile, global dairy commodity markets. This reliance on imported dairy products is likely to continue for most Asian countries, although many of them are striving towards self-sufficiency in dairy production.

There is a group of Asian countries with low per capita milk consumption and low self-sufficiency and these are likely to be the ones with most proactive dairy development programmes. These include the Philippines, Indonesia, Thailand, Malaysia, Vietnam, Cambodia and Laos.

2 Dairy farming in Asia

Dairy farming in Asia can be broadly classified into three major types of production systems:

1. *Mixed farming*, in which income from milk production constitutes only a relatively small proportion of the total farm income. Many of these farms have evolved from essentially cropping enterprises to those where livestock production is becoming more important. Milking herd sizes are generally quite small on these farms, ranging from fewer than 5 to more than 20 cows approximately.
2. *Essentially smallholder dairy farms*, where milk production has increased over recent years to become a major contributor to farm income. While in many cases, construction of the dairy facilities has evolved and more land is available, these improvements may not be sufficient to meet future requirements. Milking herd sizes are very small, generally no more than 5–10 cows.
3. *Larger specialist dairy farms*, which were established primarily to produce raw milk. Dairy facilities on these farms have been better planned to satisfy the requirements for a predetermined number of milking cows. In most cases, land would have been

© Burleigh Dodds Science Publishing Limited, 2017. All rights reserved.

allocated to produce the required fodder for the planned herd size, although in certain cases, agreements would have been made with surrounding farmers to provide the necessary forage base. Milking herd sizes on these farms would range from 20 to 100-plus cows.

The contribution of these various farming systems to the total milk produced in each country will vary with population pressures and demands for alternative land use, other than providing livestock fodder. However, Categories 1 and 2 contribute the bulk of the raw milk. The majority of dairy farmers are smallholders, with average herd sizes often as small as one to five milking cows. In the developing world, 80% of milk is produced by smallholder dairy (SHD) farmers, who thus make a significant contribution to the annual world production. Despite their high profile in the dairy industry, there are relatively few large dairy feedlots in Asian countries.

Dairy farmers around the world produce milk from six different types of ruminant animals:

- large (cattle and buffalo plus camels in Africa and yaks in Asia)
- small (goats and sheep)

Small ruminants are rarely milked in Asia. Of the two buffalo ecotypes, river buffalo are the traditional dairy stock, with swamp buffalo rarely being milked. The majority of milk in Asia is derived from cattle, with some buffalo milk produced in Myanmar, Vietnam, the Philippines and Thailand, while the large buffalo-milk-producing countries are India, Pakistan, China and Nepal.

On any dairy farm, no matter its size or location, dairy production technology can be broken down into nine key activities, which can be considered as steps in the supply chain of profitable dairy farming (Moran 2009a). Any chain is only as strong as its weakest link – thus, each step in the dairy farming supply chain must be properly managed. Weakening any one link through poor decision-making can have severe ramifications on overall farm performance and hence profits. In chronological order of their role in ensuring a profitable dairy enterprise, the 'links' are presented in Fig. 1.

It is important to note the important role of women in carrying out many of the key activities in the dairy value chain. With the cows typically being located in close proximity to the home, dairying offers more opportunities than other farming pursuits for women to become closely involved in day-to-day management. This is important in the village life in Asia, where women have traditionally been homemakers and family rearers. The cultural and religious bonds limiting their contribution to managing the family budget have frequently been loosened in many smallholder dairying communities.

In West Java for instance, Innes (1997) has documented gender roles in smallholder farm activities in four dairy cooperatives. She reported that women in the farm family were responsible for over 40% of the farm management decisions and spent 52% of their working hours on dairy-farm-related jobs. Men were largely responsible for sourcing forages, often from large distances particularly during the dry season. However, women frequently milked the cows, transported the milk to the collection centres, cleaned the shed and looked after the young stock. This has important implications for the process of technology transfer, which has traditionally been aimed at men. Since milking hygiene is largely the responsibility of women, milk quality is definitely an area where women should be targeted by extension services. Efforts should also be made to attract female participants to workshops on feeding management and young stock.

© Burleigh Dodds Science Publishing Limited, 2017. All rights reserved.

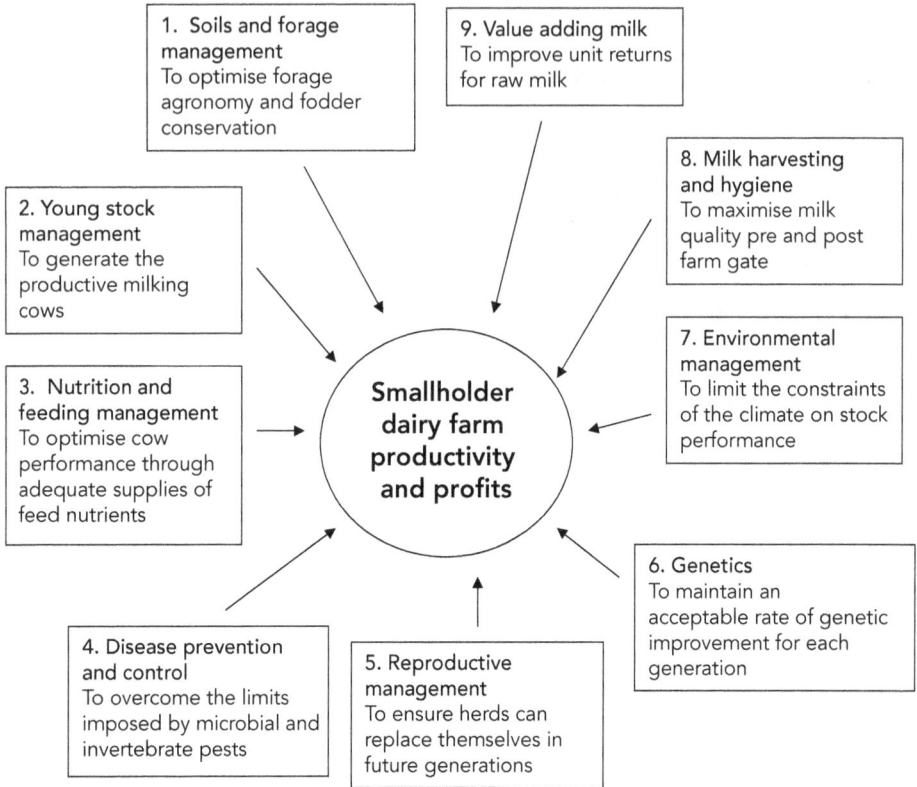

Figure 1 The nine steps in the supply chain of profitable dairy farming.

3 Supporting smallholder dairy farmers

National governments, international aid agencies and benevolent governments of, or agencies from, developed countries have devoted and are still devoting a lot of resources towards improving the productivity and profitability – hence sustainability – of the SHD industries throughout Asia. The focus is on sustainable intensification of SHD farming. The term 'intensification' requires clarification. In general terms, intensification is understood to be increases in efficiency for a unit of a given resource. For advisers and researchers of crop-livestock or pasture-based livestock production, the term is often interpreted as increasing productivity per unit of land, usually associated with an increase in stocking rate.

The national dairy development (5- or 10-year) programmes in most Asian countries concentrate much of their efforts towards the Category 2 farmers mentioned earlier, that is, smallholder dairy farms. In other words, they are trying to phase out 'part time' dairy farmers (in Category 1) and encourage 'full time' dairy farmers. National dairy plans provide government support, which often includes the establishment of dairy cooperatives.

Category 3 farmers (larger, specialist dairy farms) are usually less reliant on public support as their establishment is often financed by private investors. However, in recent years there

© Burleigh Dodds Science Publishing Limited, 2017. All rights reserved.

has been considerable interest (and investments) in larger scale, feedlot dairies. This is occurring because governments have struggled to overcome the inefficiencies of current SHD systems, such as low milk yields, poor cow fertility and high young stock mortality rates, which drastically limit their ability to greatly increase their dairy sectors to achieve self-sufficiency in dairy production.

Smallholder farms generally yield low outputs of milk per animal. However, a cost-benefit analysis can show these farms to be productive – the use of by-products or other waste as feed and multiple outputs such as calves and meat production in these farms can allow them to outshine dairying monocultures despite the latter's apparent efficiencies. Moreover, application of current technologies and a better understanding of the nutrient requirements of the animals, in addition to the requirements for growth and meat production, will lead to higher efficiency in milk and meat production.

There are many benefits that will accrue from the improved productivity and profitability of SHD farmers. In addition to higher levels of milk production (hence gross returns) per cow and/or per farm, Falvey and Chantalakhana (1999) list the following:

- year-round engagement of rural and peri-urban labour;
- utilisation of agricultural and other by-products;
- integration with cropping systems management;
- conversion of by-products into organic manure for application to crops;
- provision of nutritious and hygienic food for children;
- production of meat from male calves and older cows;
- reduction in the cost of production of meat for traditional markets as draught power declines as the primary bovine product;
- a basis for rural and peri-rural industrial development through milk factories;
- development of new products for niche exports;
- reduction in rural to urban population drift;
- draught and traction as a dairy industry by-product or adjunct; and
- landless people making a reasonable local living from dairying.

A recent industry study of SHD farming in the tropics highlights the role of SHD farming, using a SWOT analysis to evaluate the industry's strengths and weaknesses. The analysis assesses the business or industry's strengths (S), weaknesses (W), opportunities (O) and threats (T). Although Table 1, presented below, was undertaken specifically for Indonesia's SHD industry by Anon (2005), it is applicable to any SHD industry in tropical Asia. Anon (2005) then concluded that SHD farming in Indonesia, as in other tropical Asian countries,

- improves the food security of milk-producing households;
- creates employment opportunities throughout the entire dairy chain (for both producers and processors);
- is a powerful tool for reducing poverty and creating wealth in rural areas; and
- can incur relatively low production costs.

In spite of several decades of dairy farming in developing countries, the productivity of SHD farms has remained relatively low; this is due to a lack of appropriate dairy research and extension. Due to the socio-economic and agro-economic conditions of small farmers in developing countries being greatly different from those prevalent in developed countries, the farmers cannot readily adopt the science and technology available in developed

© Burleigh Dodds Science Publishing Limited, 2017. All rights reserved.

Table 1 Findings of a SWOT analysis of Indonesia's SHD industry

Components of SWOT	Findings
Strengths	Low production costs High farm income margins Low liabilities Relative resilience to rising feed prices SHD farmers are cost competitive and resilient to market fluctuations They provide a competitive source of milk supply to imported dairy products
Weaknesses	Lack of knowledge and technical skills Poor access to support services Low capital reserves and limited access to credit Low labour productivity (small herd sizes and low output per cow) Poor milk quality SHD farmers are often unable to take advantage of existing market opportunities
Opportunities	Growing demand for dairy products in developing countries Likelihood of increased milk returns Major potential to increase labour productivity Great potential to increase milk yields Employment generation Significant opportunities to improve the demand (quality and milk price) Significant opportunities to improve the supply (improving production technology)
Threats	Policy support in developed countries Exposure to competitive business forces Underinvestment in dairy chain infrastructure Unsuitable dairy development plans Environmental concerns such as a high carbon footprint Increasing consumer demand for food safety Succession of dairy farms Increasing local wage SHD rarely meets its full potential because of many threats, particularly the last four

Source: Anon. (2005).

countries. It is essential that any production technology being transferred is relevant to the needs of these smallholders as well as being feasible, given their local support networks of dairy cooperatives, advisers (government and agribusiness), creditors and milk handling and processing infrastructures. Even the most appropriate technology is rarely transferred successfully to smallholders due to a lack of effective support services. There must be institutional support to facilitate dairy industry growth through mechanisms such as provision of farmer credit, farmer training centres, well-equipped milk collection centres, processing and marketing facilities, farmer cooperatives or groups, and appropriate research and extension infrastructures and methodologies.

For intensification to be sustainable, there must then be:

- adequate infrastructure and marketing opportunities;
- access to reliable markets for increased milk production;

© Burleigh Dodds Science Publishing Limited, 2017. All rights reserved.

- promotion of dairy development through government policy;
- availability of credit for purchasing of livestock and planting pastures;
- available productive and adapted forage species;
- ready access to information;
- farm management systems that ensure adequate feed throughout the year;
- management of animal wastes;
- disease control measures; and
- adequate hygiene for milk collection.

4 Key constraints facing smallholder dairy farmers in tropical Asia

As a result of applied dairy research, development and extension over the last 20 years, Western countries have produced sophisticated dairy production systems (such as those described by Little 2012). Herd sizes have grown, efficient feeding systems have evolved and many farmers routinely monitor test results on their cows for milk production, composition and quality, and for mastitis. They then use this information for making decisions on culling milking cows and for breeding genetically improved stock. Another feature of Western dairy farming is the high degree of mechanisation, to address the high labour costs, such as milk harvesting and forage conservation systems. In addition, cheap land has allowed for grazing stock which reduces the costs of harvesting fresh forages while the lack of population pressures has allowed these farms to rapidly expand in both area and herd sizes.

Unfortunately, the dairy industries of tropical Asia have failed to keep pace with the speed of such dairy development in Western countries (Devendra 2001). It is true that the number of cows in most Asian countries has greatly increased, largely through government support for social welfare and rural development programmes, whose driving forces have been the increased demand for milk (accentuated through school milk programmes) and the concept of national food security. However, in terms of milk production per cow and feed inputs per kg of milk produced, improvements have been slow (Moran 2005, 2009a, 2012).

There are many reasons why the productivity and efficiency of SHD farming has not greatly improved over the last two decades. Many of these developing dairy industries are located in tropical regions where high temperatures and humidity and, in some cases, seasonal growing conditions, adversely affect potential milk yields. Milking cows are not well suited to the tropics because their large requirements for feed nutrients and their high internal heat production (compared to other species of livestock) cannot easily be incorporated into production systems that have to cope with poor forage quality, exposure to many disease agents and the climatic stresses that constrain cow appetite, reproductive efficiency, performance of young stock and animal health (Moran 2005).

In addition, many of the farmers, usually smallholders with fewer than 10 milking cows, have not been able to develop the skills of efficient milk production. As previously mentioned, this can be attributed more to poor extension services than to a lack of technical knowledge on tropical dairy farming. SHD farmers, who hail from regions where socio-economic and agro-economic conditions are vastly different from those in Western dairy industries, cannot readily adopt the science and technology available in developed countries. Therefore, it is essential that any production technology being transferred is

© Burleigh Dodds Science Publishing Limited, 2017. All rights reserved.

relevant to the needs of smallholders as well as being feasible, given the smallholders' local support networks of dairy cooperatives, advisers (government and agribusiness), creditors and milk handling and processing infrastructures (Devendra 2001).

Falvey and Chantalakhana (1999) categorised the factors limiting SHD production into:

- institutional factors, such as dairy cooperatives, suppliers of credit, training, extension services;
- government policies, such as development programmes, milk promotion, dairy boards;
- socio-economic factors, such as farmer education, off-farm jobs, traditional beliefs;
- technical factors, which can be further categorised into feeding, breeding, health; and
- post-farm gate factors, such as milk processing, marketing and consumption.

This analysis can be compared with a more recent study (Burrell and Moran 2004). In the early 2000s, a series of strategic planning workshops were conducted in Indonesia to identify the key constraints limiting milk production and to develop action plans to combat them. Burrell and Moran (2004) laid out the constraints and the relevant action plans by region, making separate lists for East and West Java.

In East Java, the priority industry issues and the action plans for industry development were:

1 *Low cow productivity*: improve management of feeding, reproductive management and milk harvesting.
2 *Low milk price*: reduce costs of production, improve milk quality, mediate on milk pricing, find alternative markets.
3 *Poor milk quality*: improve milking hygiene at both farm and post-farm gate, improve milk composition through better feeding management.
4 *Poor feed quality and availability*: identify better forage species (e.g. legumes), appoint quality control teams for concentrate supplies, utilise marginal land for forages.
5 *Cooperative management*: reduce management structure and merge small cooperatives, improve post-harvest technology, improve calf and heifer rearing practices.

Other industry issues were raised but were not discussed in detail. These included the need to promote fresh and manufactured dairy products, improve technology transfer, stimulate farmer motivation, work towards autonomy of cooperatives and improve collaboration between government agencies and training organisations.

West Java's priority industry issue and action plan list, developed independently, was as follows:

1 *Human resources*: improve knowledge, skills and attitudes of farmers and support staff;
2 *Poor feed quality and availability*: increase area of land for growing forages, overcome seasonality of forage supplies, reduce variability of concentrate quality;
3 *Low capital investments in industry*: invest in infrastructure for post-farm gate industry support;
4 *Small scale of farming*: increase herd sizes, overcome shortage of breeding stock;

© Burleigh Dodds Science Publishing Limited, 2017. All rights reserved.

5 *Insufficient technology*: increase supply of breeding bulls, improve feed supplies, diversify farming systems, value add milk in farming areas to help overcome farmers' low cash flows;

6 *Institutions*: improve coordination amongst service providers, introduce better control over milk quality, improve efficiency of administration in institutions.

Other industry issues were raised but were not discussed in detail. These included the need to promote fresh drinking milk, facilitate and support milk marketing and develop post-farm gate technology in milk processing.

The on-farm constraints to SHD dairy production technology in tropical Asia are many and varied. Thirty-five of the key ones were summarised by Moran (2013) and are listed in Table 2. They are categorised using the nine key activities from Fig. 1 and a range of possible solutions to overcome them are provided. An extra category 'Other on farm constraints' is included in this Table to take into account those covering farm business skills.

5 Benchmarking performance

The dynamic nature of dairy farming makes it difficult to develop a simple set of criteria with which to assess current management skills. The term Key Performance Indicators (KPI) refers to a series of measures of dairy farm performance with which realistic targets can be set after effecting improvements in feeding, herd and farm management. Such a set of KPIs for SHD farming has recently been published by Moran (2009b). All these KPIs can be quantified to provide guidelines on which ones require priority in any dairy farm improvement programme. Although some are relatively easy to quantify, others are quite difficult. Probably the simplest, and most commonly used, single measure of SHD farm performance is the average milk yield of the milking cows. The correct term for this figure is 'rolling herd average', as it is the average milk yield of all the milking cows, which on any one day will be at various stages in their lactation cycle.

This single value provides a summation of all the important aspects of SHD farm management, so any interpretation must take into account a diversity of feeding, herd and farm factors (Moran 2012). Accordingly, many dairy specialists may query its usefulness as a single measure of dairy farm performance. However, it is routinely used by farmers to describe their farm's performance in relation to their neighbour's farm and also in relation to production targets provided by many government advisers. In addition, it is often quoted by government officials when summarising the stage of development of their national dairy industries. Table 3 attempts to describe the adequacy of the farm's dairy farm management practices using the rolling herd average.

There are other factors and KPIs to consider when interpreting such data:

- differences between rolling herd averages and peak milk yields
- milk composition as an indicator of feeding management, for example:
  - Low milk fat can indicate possible subclinical rumen acidosis.
  - High milk protein can indicate good dietary energy intake.
  - However milk lactose levels are fairly constant.

© Burleigh Dodds Science Publishing Limited, 2017. All rights reserved.

Table 2 Key constraints to improved milk production on tropical Asian smallholder dairy farms and possible approaches to solutions

Key activity	Key constraints	Approaches to solutions
1. Soils and forage management	a. Low yields of forage	Use inorganic fertilisers as well as manure Reduce nitrogen volatilisation of shed effluent by directing it into water storage Optimise forage agronomy (soil preparation, weed control)
	b. Poor forage quality	Use inorganic fertilisers as well as manure Use most appropriate forage species for region Consider other forages such as tree legumes Reduce harvest intervals
	c. Shortage of dry season forages	Consider silage making of wet season forages Plan year-round forage supplies
2. Young stock management	a. High calf mortality	Better parturition management to minimise likelihood of infecting new born calf Ensure use of semen or bulls with low calf birth weights Improve colostrum feeding programme (Quantity, Quality, Quickly) Pay greater attention to navel dipping with iodine Better shed hygiene Develop skills in identifying potentially sick calves Better health management Identify causes of death or sickness and change management accordingly Improve calf housing Minimise stress in calf shed Consider feeding less milk to encourage concentrate intakes Be more aware of fluid replacers vs. antibiotics for treating calf scours
	b. Poor post-weaning growth rates	Feed adequate amounts of concentrates Ensure calf concentrates have 18% protein Feed less forages to stimulate concentrate intakes Better health management Ensure routine Clostridial vaccination programme Monitor post-weaning growth rates
	c. High wastage rates (from birth to conceiving in 2nd lactation)	Dairy cooperatives could consider heifer farms

© Burleigh Dodds Science Publishing Limited, 2017. All rights reserved.

3. Nutrition and feeding management	a. Low quality of by-products and formulated concentrates	Routine laboratory testing of ingredients and formulation Quality control during formulation Use coop system to bulk purchase quality by-products
	b. Poor performance of cows during early lactation (poor peak and daily milk yields, delayed cycling)	Ensure best forages for cows in early lactation, never rice straw Ensure enough forages are fed (30–50 kg fresh grass per cow per day) Monitor total dry matter intakes and increase if insufficient Consider wilting fresh forages to stimulate intake Ensure at least 16% protein in total ration Ensure all feeds are palatable Ensure adequate clean drinking water Provide Ca & P supplements in formulation Check if sufficient rumen buffers in concentrates Do not make concentrates and water into a slurry Chop forages to reduce selection and wastage Address any heat stress issues
	c. Cows (particularly high genetic merit cows) do not cycle for many weeks after calving	Ensure sufficient forages and concentrates are fed Check to see if rapid loss in weight or body condition Ensure at least 16% protein in total ration Consider vet checking for ovarian or uterine health
	d. Seasonality of milk production	Plan year-round sourcing (growing or purchasing) of quality forages Ensure year-round supplies of by-products and formulated concentrates Ensure adequate supplies of drinking (and washing) water throughout the dry season Ensure adequate cow comfort throughout the year
	e. Little profit in milking cows	Check milk income less feed costs (MIFC) Be aware of marginal milk responses if feeding too much Set realistic target milk yields and feed to achieve them Ensure ration is balanced for nutrient contents Maybe feeding too many cows for available feed supplies Feed fewer cows better
4. Disease prevention and control	a. Problems with lameness	Check floors for ease of walking on them Consider foot bath for all stock Check ration if too much concentrates causing laminitis Undertake locomotion test and treat affected cows

(continued)

© Burleigh Dodds Science Publishing Limited, 2017. All rights reserved.

Table 2 Continued

Key activity	Key constraints	Approaches to solutions
	b. Problems with mastitis	Identify subclinical cases with California Mastitis Test
		Ensure one towel to wash only one cow policy
		Treat every infected cow with antibiotics ensuring withdrawal period is followed
		Milk-infected cows last
		Initiate routine dry cow antibiotic therapy
		Consider culling chronically infected cows
	c. High calf and heifer morbidity and mortality	Follow procedures as in young stock management
	d. General animal health problems	Develop skills in identifying potentially sick stock
		Routinely inspect stock for external parasites
		Isolate sick stock
		Improve routine use of vaccinations
		Routinely use quality and viable pharmaceuticals
		Reduce the degree of exposure by improving shed hygiene
		Consider testing for internal parasite egg counts
		Reduce any overuse of antibiotics
		Find better-trained veterinarians
5. Reproductive management	a. High age at first calving	Follow procedures for poor post-weaning growth rates in young stock management
	b. Low 100 day in calf rate (pregnant within 100 days from calving) or high 200 not in calf rate (not pregnant within 200 days of calving)	Better feeding management during early lactation
		Check AI (artificial insemination) techniques
		Can veterinarian confidently undertake pregnancy diagnosis?
		Pay closer attention to heat detection
	c. High number of services per conception	Improve AI techniques or check that technician is sufficiently skilled
		Pay closer attention to heat detection
		Consider vet checking for ovarian or uterine health
	d. Low % mature cows are milking	This is a simpler measure of poor reproductive performance so follow procedures above
	e. Increasing the proportion of heifer calves	There may well be a role for sexed semen in well-managed dairy farms

© Burleigh Dodds Science Publishing Limited, 2017. All rights reserved.

6. Genetics	a. Poor milking cow quality	This generally is not an issue because the genetic merit of imported dairy heifers is likely to be better than any cow on the farm It is quite likely that the performance of most milking cows will be limited by the environment (feeding, disease, heat stress, etc.) rather than genetic merit Be aware of the genotype by environment interaction which means that high genetic merit stock require better levels of feeding and farm management to express their higher potential performance
	b. Most suitable genotype for the system	Some countries will not allow Jersey crossbreds to be imported hence the imported Friesians limit the dairy production to the highlands If Jerseys are allowed to be imported, they may well prove the more profitable breed in lowland regions
	c. Difficulty of collecting robust data from genetic improvement programmes	More emphasis on permanent identification of heifers Pay greater attention to maintaining cows in milking herds for relatively lengthy periods
7. Environmental management	a. High incidence of heat stress during the 24-hour period	Count respiration rates to quantify degree of heat stress Pay closer attention to heat dissipation - Check shed design for ventilation - Consider artificial cooling (sprinklers and fans) Feed cows during the evening, when cooler Consider outside area for night time cooling and heat (cycling) observations Feed better quality forages to reduce internal heat production
	b. High incidence of animal health problems due to poor shed hygiene	Improve shed hygiene Remove manure more frequently Isolate sick stock
	c. Reduced forage quality due to high temperatures and rainfall	Unfortunately it is not easy since tropical forages are more fibrous than temperate forages. Soil testing can assist with overcoming monitoring leaching due to high rainfall
8. Milk harvesting and hygiene	a. Poor milk composition (fat and protein content)	Address any limiting feed nutrient deficiencies Ensure sufficient forage intake to maintain milk fat content Maximise cow comfort so cows will maintain their appetite

(continued)

© Burleigh Dodds Science Publishing Limited, 2017. All rights reserved.

Table 2 Continued

Key activity	Key constraints	Approaches to solutions
	b. Poor milk quality (bacterial contamination)	Improve milking hygiene (hot water, detergent, sanitiser) Ensure machine milkers are operating effectively (short milking times, correct pulsation rate) Ensure rubber liners are correctly replaced Address any mastitis problems Ensure rapid milk cooling Could be a post-farm gate issue hence outside farmer's control
9. Value adding milk	a. Poor milk returns	Consider value adding to improve unit milk returns
10. Other on-farm constraints	a. The small farm size restricts development potential	Dairy cooperatives could develop cow colonies (see next section)
	b. Poor profitability of dairy farming	Quantify profitability over 6–12 month period Quantify milk returns and overall farm income (actual and potential) Quantify cost of production (COP) Be aware that increased profitability can result from decreased COP as well as increased farm income Dilute fixed costs with higher farm cash throughput
	c. Low capital resources for investing in farm infrastructure	Seek alternative low interest loans
	d. Poor dairy farming skills	Institutional support to improve farmer training
	e. Underdeveloped entrepreneurial skills in dairy farmers	Work closely with potentially successful farmers to help develop these skills Provide training in farm business management and developing farmer business skills
	f. Poor farmer–management dairy coop relationships	Become more vocal to improve them

© Burleigh Dodds Science Publishing Limited, 2017. All rights reserved.

Table 3 Interpreting the adequacy of dairy farm management from cow milk yields: range in average herd milk yields on tropical Asian dairy farms

Herd milk yield (kg/cow/day)	Adequacy of dairy farm management practices
5	Very poor feeding and herd management and low genetic merit cows (or milking buffalo)
7	
9	Typical of many SE Asian smallholder farms, even with high-grade Friesians
11	Gradual response with grade and crossbred Friesian cows to improved feeding, herd, young stock and shed management.
13	*Milk yields of 15 kg/day are considered acceptable by many government dairy*
15	*advisers.*
17	
19	
20	Potential level in **lowland humid tropics** following improved management of body condition throughout lactation
25	High genetic merit cows in **tropical highlands** or **lowland dry tropics** with excellent farm management
30	Typical peak milk yields in herds with 25 kg/cow/day rolling herd averages
35	Unrealistic in SE Asia except where all major constraints to milk production have been overcome

- Excessive body condition, as it is indicative of low protein diets. This results in:
  - inability of cow to partition nutrients from body reserves to milk synthesis and
  - poor fertility as cows cannot easily cycle, hence conceive.
- Very poor body condition, as it is indicative of low energy intake. It should be noted that:
  - high genetic merit cows preferentially partition body reserves to milk synthesis and
  - cows will not cycle due to excessive weight loss.
- Herd dynamics, as it can indicate adequacy of dairy farm management. In particular:
  - excessive number of dry non-pregnant cows can indicate very poor farm management and
  - low percentage of lactating adult cows can indicate poor farm management.

6 Case study: cow colonies

In many tropical Asian countries, considerable attention has been paid to large-scale investments in 'cow colonies'. These consist of large dairy sheds, holding 50 or more cows, that are owned by a number of SHD farmers, and surrounding large areas that are meant for forage production. Although smallholders still own and manage their own herds in these large sheds, the perceived benefits of cow colonies lie in the magnitude of size of the total herd management. While such an approach can overcome many constraints to production, it may introduce other constraints.

© Burleigh Dodds Science Publishing Limited, 2017. All rights reserved.

The potential benefits of cow colonies are as follows:

- greater investment potential since cooperatives have more borrowing power than individual farmers;
- use of mechanical forage choppers and milking machines;
- presence of contract labour to rear young stock;
- growth of large areas for forages, such as maize;
- less wastage in recycling manure to forage production area, through building effluent ponds to minimise volatilisation of nitrogen from urine;
- bulk handling of conserved forages using large-scale silage bunkers;
- easier communication between advisers and farmers and between farmers themselves;
- easier implementation of training programmes involving practical skills as well as technical theory;
- easier monitoring of post-training application of new skills;
- better motivation of farmers to improve management practices;
- easier monitoring of individual farmer's milking hygiene practices and hence individual remuneration for better quality milk;
- the concentration of farmers in one place, which provides an ideal opportunity to introduce other motivational techniques such as regular awards for best management practices;
- better coordination of forage production, cow feeding, insemination, animal health, milk handling, etc.;
- training of farmers in specialist skills such as machine milking or calf rearing;
- the installation of cooling units on site;
- more rapid cooling of milk and greater availability of hot water for more effective cleaning and sanitising equipment;
- increased likelihood of sufficient milk production to justify small value adding operations to benefit small dairy cooperatives; and
- greater potential returns to the local dairy cooperative, hence the farmers themselves.

Unfortunately, these impressive facilities have gone hand in hand with high-profile projects – for instance, these facilities have been stocked with imported pregnant Friesian heifers. The high mortality rates so far experienced in countries such as Indonesia suggest that the colony feeding and herd management practices have yet to be improved to benefit from these high genetic merit animals.

The following are problems that have often been associated with cow colonies:

- the sheds are constructed and filled with cows before the forage production area has been developed, leading to many poorly fed cows;
- insufficient attention is placed on growing out non-revenue generating, young stock;
- poorly planned forage production areas – for example, minimal water is available for irrigation during the dry season;
- insufficient land allocated to forage production, partly because of provision of insufficient daily forage allocations to achieve target milk yields;
- incorrect perception that rice straw, sugar cane tops and over mature maize stover are suitable forage sources for milking cows, particularly when target milk yields are 15 l/cow/day or more;
- lack of understanding of the potential of forage and tree legumes as important forage sources for high-yielding cows;

© Burleigh Dodds Science Publishing Limited, 2017. All rights reserved.

- potential spread of disease because of variable management practices of individual farmers during calf rearing – for example, the spread of mastitis when milking machines are used by farmers;
- poor understanding of the need for milking hygiene when using milking machines – for example, the need to regularly replace milk liners and to test machine performance;
- continual breakdown of machinery, choppers and milking machines;
- lack of highly trained and well-skilled labour for year-round supply of quality forages;
- failure of senior managers to develop both short-term and long-term views on development programmes;
- difficulties involved in regularly sourcing finances for completion of these large-scale capital development projects, such as provision of milking equipment, durable forage choppers;
- skilled individuals have limited responsibility as the management teams are small – the larger the operation the more essential that skilled individuals be given more responsibility in specialist areas, such as forage production, animal health, milk quality;
- the expectation on management teams of large-scale cow colonies to oversee nearby smallholder farms;
- failure of senior managers to find and keep quality staff with capabilities of solving both day-to-day small management problems as well as contribute to large-scale development. This problem could be addressed by employing bright young animal science graduates who would be prepared to live as well as work in villages near cow colonies. With the penalties imposed by milk processors, returns on these large capital investments are markedly reduced because of the low unit milk returns through poor quality milk. Small investments, such as steam cleaners and small hot water units, become even more effective in light of the large capital costs of sheds, silage bunkers, etc.; and
- too many cows. It is more profitable, as has been found true in smallholder ventures, to 'feed fewer cows better'.

Poorly resourced SHD farmers, whose businesses are often in 'survival mode', can become very individualistic and can take time to develop the cooperative, sharing nature required for successful cow colonies. This has been given as a common reason for their poor success rate in countries with relatively new SHD industries such as Indonesia. The problems associated with cow colonies show the need to take a holistic view that accounts for each step in the dairy value chain.

7 Summary and future trends

After several decades of dairy development in many Asian countries, typical milk yields per cow per day still range between 8–10 kg as compared to average yields of 20–30 kg in developed countries. In addition, the average calving interval of dairy cows on SHD farms is commonly as long as 16–20 months, when it could be reduced to 14–15 months. With regard to young stock management, heifer ages at first calving are more commonly 30–36 months rather than the 24–28 months commonly found in temperate, more developed dairy industries. This clearly shows the low levels of farm productivity in tropical Asia. Many technical solutions are available (as in Table 2), but they must be carefully selected so they will be suitable for small farmers and their

© Burleigh Dodds Science Publishing Limited, 2017. All rights reserved.

socio-economic conditions. This means that scientists and extension workers must be able to understand factors influencing acceptance when transferring such technology to farmers. Scientific knowledge alone cannot solve small-scale farm problems (Falvey and Chantalakhana 1999).

Policy makers should resist the all too common assumption that development efforts should move from smallholders towards supporting larger scale, 'more efficient' milk producers to meet growing consumer demand. Instead, growing demand should be used as a stimulus to help continue and sustain SHD enterprises particularly when they face increasing barriers to participation in value chain markets (Ahuja et al. 2012).

If well organised, SHD can compete with large-scale, capital intensive 'high tech' dairy farming systems as practised in both developed and developing countries. However, SHD development plans must include strategies to increase competitiveness in all segments of the dairy industry chain, namely input supply, milk production, processing, distribution and retailing (APHCA 2008; Otto et al. 2012). The future for SHD farming in tropical Asia is optimistic so long as the industry can rectify many of the constraints to improving domestic production of raw milk, particularly those at the farm level.

8 Where to look for further information

A standard introduction to the subject is J. B. Moran, *Tropical Dairy Farming* (see Moran 2005 in the References and further reading section for full details).

The best single source of information on smallholder dairying in Asia is the Asia Dairy Network jointly established by the FAO and the Animal Production and Health Commission for Asia and the Pacific (APHCA) (http://www.dairyasia.org/). The site includes information resources and key contacts.

Centres of expertise include:

- The International Livestock Research Institute (ILRI) (http://asia.ilri.org/)
- The National Dairy Research Institute in India (http://www.ndri.res.in/)
- The author's own consultancy, which has undertaken numerous projects to support smallholder dairy farmers in countries such as Indonesia, Malaysia, Thailand, Bangladesh, Myanmar and India (http://www.profitabledairysystems.com.au).

There are a number of current research projects designed to support dairy farmers in Asia and which identify current problems and ways of tackling them, including:

- The CGIAR's Research Program on Livestock and Fish which includes improving the dairy value chain for Indian smallholders (http://livestockfish.cgiar.org; an overview can be found in Rao et al. (2014))
- The Smallholder Dairy Development Programme coordinated by the FAO and others, focusing on Bangladesh, Myanmar and Thailand (http://www.dairyasia.org/ projects)
- The Market Access for Smallholder Farmers (MASF) Project, coordinated by Practical Action, which supports dairy farmers in Nepal (http://practicalaction.org/ region_nepal_masf_project)

© Burleigh Dodds Science Publishing Limited, 2017. All rights reserved.

9 References and further reading

Ahuja, V. B., Dugdill, N., Morgan, N. and Tiensin, T. (2012). Smallholder dairy development in Asia and the Pacific. In: Planning dairy development programs in Asia. *Proceedings of Symposium 15th AAAP Congress.* Bangkok, Thailand. pp. 77–85. Available from: http://www.dairyasia.org/file/Proceedings_dairy.pdf

Anon. (2005). Indonesia's dairy farming industry SWOT analysis 2005. Stanton, Emms and Sia, Singapore, September 2005.

Anon. (2014). *Dairy Asia: Towards Sustainability (Proceedings of an international consultation held in Bangkok, Thailand 21-23 May 2014),* FAO Regional Office for Asia and the Pacific, Thailand.

APHCA (2008). Developing an Asian regional strategy for sustainable small holder dairy development. *Proceedings of FAO/APHCA/CFC funded workshop,* Chiang Mai, February 2008.

Burrell, D. E. and Moran, J. B. (2004). Developing a strategic plan for Indonesia's small holder dairy industry. In H. K. Wong, J. B. Liang, Z. A. Jelan, Y. W. Ho, Y. M. Goh, J. M. Panandam and W. Z. Mohamad (eds). *Proceedings of the 11th Animal Science Congress.* Vol. 1. pp. 143–6. September 2004. Asian-Australasian Association of Animal Production Societies, Malaysia.

Delgardo, C., Rosegrant, M. and Wada, N. (2003). Meating and milking global demand: Stakes for small-scale farmers in developing countries. The livestock revolution. A pathway to poverty? *ATSE Crawford Fund Conference.* Canberra, Australia. pp. 13–23.

Devendra, C. (2001). Small holder dairy production systems in developing countries: Characteristics, potential and opportunities for improvement. *Asian-Australian Journal of Animal Science.* 14. pp. 104–15.

Falvey, L. and Chantalakhana, C. (1999). *Small Holder Dairying in the Tropics.* Kenya: ILRI.

Falvey, L. and Chantalakhana, C. (2001). Supporting smallholder dairying in Asia. *Asia-Pacific Development Journal.* 8 (2). pp. 89–99.

Hemme, T. and Otto, J. (2010). *Status and Prospects for Smallholder Milk Production: A Global Perspective.* Rome: FAO.

Innes, T. (1997). *The Role of Women in Dairy Farming. Baseline Data Collected from Four Cooperatives in West Java, Indonesia.* Ottawa: Canadian Cooperative Association.

Little, S. (2012). Feeding systems used by Australian dairy farmers. Grains2Milk. Dairy Australia website. Available from: http://www.dairyaustralia.com.au/-/media/Documents/Animals%20feed%20and%20environment/Feed%20and%20nutrition/Feeding%20Systems%20latest?Aus%20five%20main%20feeding%20systems.pdf

Moran, J. B. (2005). *Tropical Dairy Farming. Feeding Management for Small Holder Dairy Farmers in the Humid Tropics.* Melbourne: CSIRO Publications. Available from: http://www.publish.csiro.au/nid/197/issue/3363.htm

Moran, J. B. (2009a). *Business Management for Tropical Dairy Farmers.* Melbourne: CSIRO Publishing. Available from: http://www.publish.csiro.au/nid/220/issue/5522.htm

Moran, J. B. (2009b). Key performance indicator's to diagnose poor farm performance and profitability of smallholder dairy farmers in Asia. *Asian-Australian Journal of Animal Science.* 22. pp. 1709–17.

Moran, J. B. (2012). *Managing High Grade Dairy Cows in the Tropics.* Melbourne: CSIRO Publishing. Available from: http://www.publish.csiro.au/nid/220/issue/6812.htm

Moran, J. B. (2013). Addressing the key constraints to increasing milk production from small holder dairy farms in tropical Asia. *International Journal of Agriculture* and *Biosciences.* 2 (3). pp. 90–8.

Otto, J., Costales, A., Dijkman, J., Pica-Ciamarra, U., Robinson, T., Ahuja, V., Ly, C. and Roland-Holst, D. (2012). Livestock sector development for poverty reduction: An economic and policy perspective. Livestock's many virtues. FAO. Available from: http://aphca.org/Avian%20Influenza%20Alert/i2744c00.pdf

Rao, C. K. et al. (2014). *Smallholder Dairy Value Chain Development in India and Selected States (Assam and Bihar): Situation Analysis and Trends.* Kenya: International Livestock Research Institute (ILRI).

Thappa, G. (2009). *Smallholder Farming in Transforming the Economies of Asia and the Pacific.* Rome: International Fund for Agricultural Development (IFAD).

© Burleigh Dodds Science Publishing Limited, 2017. All rights reserved.

Chapter 4

Improving smallholder dairy farming in Africa

J. M. K. Ojango, R. Mrode, A. M. Okeyo, International Livestock Research Institute (ILRI), Kenya; J. E. O. Rege, Emerge-Africa, Kenya; M. G. G. Chagunda, Scotland's Rural College (SRUC), UK; and D. R. Kugonza, Makerere University, Uganda

1 Introduction

Africa hosts an estimated 310 million head of cattle, representing 20.9% of the world cattle population (FAOSTAT, 2014). The continent, however, produces a relatively small proportion (5.8% in 2013; FAOSTAT, 2016) of the global milk from cattle. It is estimated that 80% of the milk produced in Africa is from smallholder dairy production systems (FAO, 2016). Smallholder dairy production systems are defined as systems where less than 10 head of cattle are reared on land sizes that vary from less than 0.2 hectares to 4 hectares. Smallholder livestock keepers represent an estimated 20% of the world population and farm most of the agricultural land in tropical areas (McDermott et al., 2010). Within the smallholder systems in Africa, dairy production is practised under very different circumstances depending on climatic variability between regions, availability of feed and land resources, the economic ability of the producers to access the production resources as well as consumer demands and available markets (Peeler and Omore, 1997; Devendra, 2001b; Thornton et al., 2007; Banda et al., 2012; Marshall et al., 2015). The

http://dx.doi.org/10.19103/AS.2016.0005.38
© Burleigh Dodds Science Publishing Limited, 2017. All rights reserved.

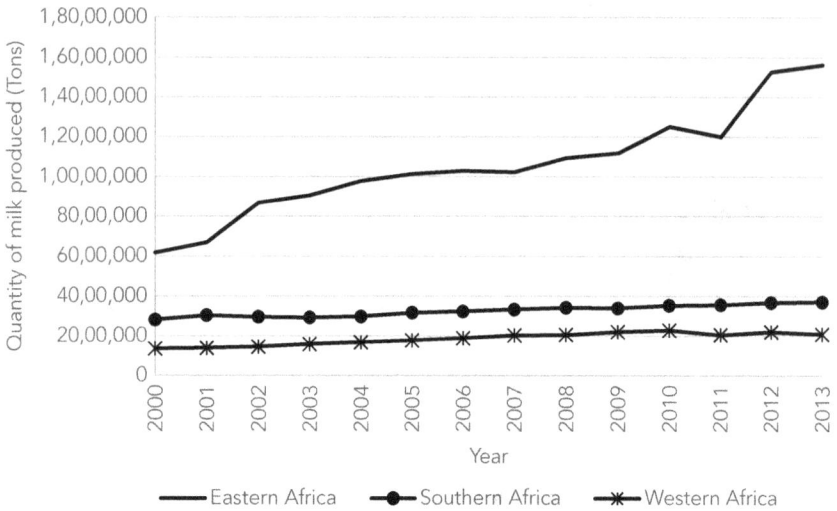

Figure 1 Differences in annual milk production from different regions of sub-Saharan Africa over time (FAOSTAT, 2016).

variability of production systems across regions is further reflected in differential quantities of milk production from cattle over time as illustrated in Fig. 1.

Dairy production under smallholder systems is considered to be a market-oriented enterprise, contributing to food and nutritional security of communities and ensuring a regular income for the farming households. Dairy animals also play a central role in socio-economic activities of many households, while the practice of dairying provides direct and indirect employment to a large number of people. It should, however, be noted that farmers value both marketable and non-marketable by-products of their animals, consume part of the produce themselves, and appreciate intangible benefits of the animals in insurance, financing and display of status within societies (Moll et al., 2007).

The economic impact of dairy production is not confined to individual households. Rural wage rates and the opportunity costs of family labour are greatly influenced by costs of dairy production under smallholder farming systems (Staal, 2001). The impact of smallholder dairy production on rural economies has resulted in substantial development support for the enterprises from both national and international agencies (Chagunda et al., 2015), including the Bill and Melinda Gates foundation (BMGF), the United States Agency for International Development (USAID), the International Fund for Agricultural Development (IFAD), Stichting Nederlandse Vrijwilligers Netherlands (SNV), Heifer International (HI) and the Natural Resources Institute Finland (Luke).

This chapter presents a general overview of existing smallholder dairy production systems and management practices in sub-Saharan Africa (SSA), highlights key challenges and opportunities in the systems, and presents intervention options for sustainable change.

© Burleigh Dodds Science Publishing Limited, 2017. All rights reserved.

2 Sub-Saharan Africa

2.1 Classifying smallholder systems in SSA

There is no single system for classifying dairy production systems that can be mapped across all the regions of SSA. A review on classifications and definitions of livestock production systems by Robinson et al. (2011) highlighted the need for improvement in the classification and mapping of systems. Classification methods used in various studies include definitions that use combinations of resources available, and data-driven definitions based on statistical methods that involve clustering of spatial units by cattle and human population densities, cultivation intensity, livestock management practices and the elevation on the area (Mburu et al., 2007). Statistical groupings are, however, sensitive to both geographical region and value range, and hence cannot be systematically replicated. Major livestock production systems classified through a typology that integrates natural resource potential, population density and market access developed by Herrero et al. (2009, 2010) are presented in Box 1.

Box 1 Classification of livestock production systems

Agro-pastoral and pastoral systems characterized by low population densities, low agro-ecological potential and weak linkages to markets. Crop production is marginal and livestock predominate as a source of livelihood.

Extensive mixed crop–livestock systems characterized by rain-fed agriculture, medium population densities, moderate agro-ecological potential and weak linkages to market. Farming practices incorporate crop and livestock with limited use of purchased inputs.

Intensive mixed crop–livestock systems characterized by high population densities, irrigation or high agro-ecological potential and good linkages to markets. Farming practices incorporate crops and livestock, but with intensive use of purchased inputs.

Industrial systems characterized by large vertically integrated production units and in which feed, genetics and health inputs are combined in controlled environments.

Source: Herrero et al. (2009, 2010).

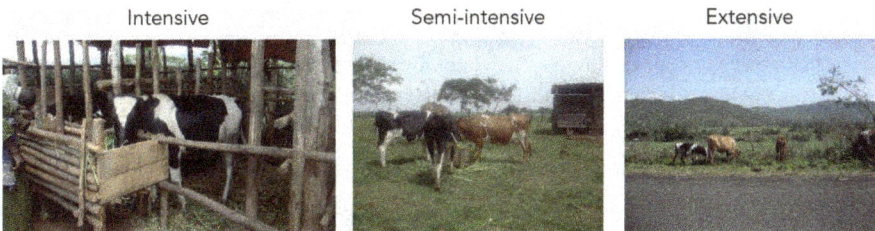

Intensive   Semi-intensive   Extensive

Figure 2 Animals reared under different smallholder management systems.

© Burleigh Dodds Science Publishing Limited, 2017. All rights reserved.

Table 1 Cattle management systems adopted for smallholder dairy production in Africa

Production system	Main characteristics of cattle management	Main cattle feeds provided	Breeds/genotypes of animals reared
Intensive mixed crop–livestock systems (often in urban to peri-urban areas)	• Stall feeding (zero grazing): all feeds required brought to animals • Good attention to preventive health care of animals • Water availed to the cows in the unit • Labour intensive • Animals produce 8–20 litres of milk per day	• Planted mixed grasses and fodders • Good quality crop residues • Supplementation with commercially manufactured concentrate feeds	• Exotic breed types (Holstein Friesian, Jersey, Ayrshire, Guernsey, Brown Swiss) • High-grade exotic crosses with local breed types
Semi-intensive mixed crop–livestock systems	• Some stall feeding and some grazing: animals confined in paddocks with some feed provided in troughs • Generally good attention to well-being and healthcare of animals • Water brought to the paddocks • Animals produce 7–15 litres of milk per day	• Planted grasses • Crop residues • Little supplementation with commercially manufactured concentrate feeds	• Exotic breed types • High- and medium-grade exotic crosses with local breed types
Extensive mixed crop–livestock systems	• Animals grazed on natural pastures often on communal land (roadsides, beside river beds) • Poor disease control: Animals treated as diseases occur • Water available from rivers and ponds • Animals produce 2–5 litres of milk per day	• Natural pastures • Crop residues; cows allowed to graze on harvested lands • Little or no supplementation with concentrates	• Medium- to low-grade exotic crosses • Local Indigenous breed types

A majority of the smallholder dairy production is carried out in crop-dairy systems which benefit from the synergies between the dairy and the crop enterprises. Examples of these systems are presented in Table 1 and illustrated in Fig. 2.

2.2 Dairy cattle breeds reared in smallholder systems

The breeds reared for dairy production under smallholder systems differ across the regions of SSA as the farmers keep a multiplicity of crosses between local (*Bos indicus*) and imported exotic (*Bos taurus*) animals. Over time, smallholder farmers cross-breed and

© Burleigh Dodds Science Publishing Limited, 2017. All rights reserved.

Box 2 Categories of livestock genotypes found in developing countries

- *Indigenous genotypes*: Tropically adapted breeds unique to Africa and/or Asia. Play critical roles in the socio-economic and cultural orientation of the communities raising them. Diversity of indigenous Zebu breeds found in Africa are presented in Rege and Tawah (1999) and Rege et al. (2001).
- *Exotic genotypes*: Breeds developed to produce high quantities of a specific product that have been introduced into the target systems to improve livestock productivity, usually from the developed world – mainly Europe and North America (e.g. Holstein-Friesian, Jersey, Ayrshire, Guernsey, Brown Swiss); also breeds introduced from other developing regions/ countries, e.g. the Boran originating from Kenya/Ethiopia and introduced in southern Africa or Sahiwal originating from Pakistan/India and now reared in Kenya; usually single product focus (e.g. Holstein–Friesian cattle), but also dual-purpose (e.g. Sahiwal cattle). The most predominant exotic dairy cattle breed type across all regions is the Holstein Friesian.
- *Cross-breeds*: Animals derived from cross-breeding either indigenous and exotic breeds or two different exotic breeds. A wide range of these exist, from those having mostly indigenous blood to those nearly exotic in their composition. A sample listing of cross-bred cattle in developing countries with published information on their productivity is presented in Galukande et al. (2013).
- *Synthetic / composite breeds*: Animals derived from systematic cross-breeding involving three or more breeds (usually exotic and indigenous) followed by generations of inter se mating to achieve stabilization.

Source: Adapted from Rege et al. (2011).

replace their animals in response to their needs, thus creating a mosaic of genotypes in the systems (Galukande et al., 2013; Leroy et al., 2015; Roschinsky et al., 2015). The main categories of dairy genotypes found in SSA are presented in Box 2.

Numbers of the different breed types and cross-bred combinations are not known. Countries in SSA have no regular mechanisms for collecting livestock population information through census, and more than 80% of the breed populations have no recorded data over the last ten years (FAO, 2015). Comprehensive definitions of breeds to distinguish different populations and descriptions of the production environments under which the breeds are reared are also scarce (FAO; 2007, 2015).

3 Management practices in smallholder dairy systems

Smallholder farmers tend to adapt models of cattle management in line with the resources at hand in a bid to meet the market demands for dairy products while at the same time balancing the social and cultural values associated with owning cattle. Husbandry and management skills of the farmers are reflected through the choice of breeds that they keep and level of animal care meted within a given production system. An understanding of the general management practices related to feed, water, animal health, handling

© Burleigh Dodds Science Publishing Limited, 2017. All rights reserved.

and breeding in the smallholder systems and the socio-economic factors that influence management decisions is a prerequisite to unlocking the potential of dairy animals within the systems. Practices adopted tend to be outlined as constraints to dairy production in developing countries in published literature (Devendra, 2001a; Nkya et al., 2007; Tebug et al., 2012; Makoni et al., 2013; SNV, 2013; Marshall et al., 2015).

3.1 Feed, fodder and water

It is estimated that smallholder dairy farmers can produce up to 70% of the feed required from their own resources. In the intensive zero-grazing production systems, for example, fodders for animals when not available or in short supply on own farms or on rented land are either collected from public land or purchased from other farmers. In dry seasons, feed resources are scarce. The quality of forages thus greatly fluctuates over the lactation cycle of the animals (Reynolds et al., 1996; Banda et al., 2012; SNV, 2013). Availability of animal feed is identified as one of the greatest constraints to improving dairy productivity within smallholder farming systems of SSA.

Actual quantities of feeds provided to animals in smallholder systems are not documented, and the quality of feed provided is variable (Reynolds et al., 1996; Banda et al., 2012). Zero-grazed animals in these systems may not get sufficient feeds of the right quality (Msangi et al., 2004; Banda et al., 2012). Nutrient deficiencies result in a negative energy balance, which in turn contributes to low fertility of animals exhibited through a high rate of repeat services when artificial insemination (AI) is used (Nkya et al., 2007). Animals with constrained energy reserves tend to have a delayed onset of their next reproductive cycle and, if served, are not able to carry pregnancies to term.

Calves in all the systems are either fed milk using buckets or left to suckle one milking quarter of their mothers' udder after milking the other quarters (Nkya et al., 2007). Many farmers, however, do not leave enough milk in the udder for the calves. Calf feed supplementation is rarely practised, resulting in malnutrition, retarded growth and delayed puberty. Reynolds et al. (1996) indicated that the survival rate of calves in smallholder systems is greatly affected by the value placed on them by the farmers. In the more intensive production systems, female calves receive better care than male calves, as females are viewed as potential replacement animals. Male calves could be raised for sale as beef animals; however, the market price for meat in many countries does not justify rearing male calves on milk to produce quality products such as veal.

Smallholder farmers in most countries rely on seasonally available sources of water for their dairy enterprises. These include rivers, wells and rain water. Very limited investments are made in water harvesting during periods of high rainfall, its storage, and also the use of ground water. Water scarcity limits development and adoption of technologies in the mixed crop–livestock systems. Provision of water for animals in many smallholder farms is rationed in relation to its availability, thus negating the benefits of investment in breeds for potentially high milk production (Forbes and Kepe, 2014).

3.2 Animal health and disease control

Vector borne diseases, notably East coast fever spread by ticks, trypanosomiasis spread by tsetse flies and anaemia caused by worm infestations, negatively impact the potential for increased dairy productivity in vast areas of SSA (Nkya et al., 2007). Increasing amounts of information on diseases affecting cattle in Africa, and various options for treatment and the

© Burleigh Dodds Science Publishing Limited, 2017. All rights reserved.

management of animal health, are accessible through various web resources (South Africa, 2014; ILRI, 2016). Parasitic diseases cause serious losses in dairy productivity through both mortality and morbidity of animals in smallholder farming systems. Mortality rates (up to 30%) have been reported for young Holstein Friesian animals raised under good management systems in Kenya (Menjo et al., 2009), whereas morbidity rate of 62% and mortality rate of 22% have been reported for calves within their first six months of life under smallholder systems in Ethiopia (Wudu et al., 2008). Within smallholder farms, the situation is more serious due to inadequate access to appropriate disease control measures.

Diseases related to production and management of animals such as mastitis, foot and leg problems, as well as reproduction- and feed-associated disorders, are a challenge in many smallholder farms (Tebug et al., 2012). Mastitis is one of the most common production diseases in smallholder dairy systems, resulting in great economic losses (Chagunda et al., 2015). Ultimately the control of diseases at farm level is the responsibility of the farmer. With support from national systems and private sector actors in service provision, most critical diseases of animals in SSA could be kept at bay. However, in most countries, legislations and regulations related to animal disease control, veterinary supplies and animal health services are weak. This has resulted in the haphazard use of various products and in the emergence of drug-resistant strains of disease-causing organisms (Tebug et al., 2012). Farmer training and innovative ways of regulating and monitoring use of antibiotics at farm level are required.

Community-led animal health strategies such as vaccination programmes driven by farmer groups and executed by private veterinarians and community animal health workers, and community-based disease and vector control (e.g. community dip-tanks and community-coordinated rotational grazing) could greatly benefit smallholder dairy farmers in Africa (Rege et al., 2011).

3.3 Animal handling structures and living environment

Structures to facilitate animal management such as housing, milking facilities, animal holding crushes, troughs for feed and water, and disposal of manure are highly variable. In intensive zero-grazing systems, cows are kept in pens throughout the year, and feed provided for them. The animals are housed in structures made of various construction materials. Roofs may be of thatch grass, iron sheets or a combination of plastic sheets and grass thatch, whereas floors may be of mud, concrete or sand. Bedding may or may not be provided for the animals. Stalls are often made from wood, and are poorly drained, resulting in accumulation of slurry, notably in the rainy seasons. This makes it difficult to keep the stalls clean, and is detrimental to the animals as the slurry provides a medium for proliferation of pathogens.

Animals under semi-intensive smallholder systems tend to be in better environmental conditions as they have more space available without the accumulation of slurry. Under these conditions, animals are better able to balance their nutrient intake through both grazing and stall feeding.

Extensive grazing does not lend to improved management as smallholder farmers using this system depend on communal grazing areas with low forage quality and limited quantity due to overgrazing typified as 'a tragedy of the commons'. Animals thus spend high amounts of energy moving in search of feed. Farmers in these systems also rely heavily on crop by-products and dry mature grass of poor quality for feeding their animals during the dry season.

© Burleigh Dodds Science Publishing Limited, 2017. All rights reserved.

Adaptation of management techniques and mechanization of processes through introduction of portable low energy equipment such as milking machines, chaff cutters and tractor carts to carry feeds and fodder would reduce the labour requirements within the smallholder systems. However, for capital inputs to be cost-effective, greater volumes of production are required (Staal, 2001).

3.4 Husbandry and breeding management

Husbandry and breeding management of animals in smallholder farming systems is variable. Animals reared are reported to have long calving intervals (up to 18 months). Practices related to timing intervals for re-breeding animals with high milk production potential under smallholder farming conditions are directly adopted from those used under systems where feed and water resources are not limited. Measures of reproductive efficiency and understanding how to manage introduced breeds are often not clear to the farmers, leading to inefficient reproductive management. Low nutrient availability and environmental factors such as diseases, high ambient temperatures and the housing environment for high-yielding cows significantly impact their milk production and reproductive performance (Walshe et al., 2011). Strategic management and monitoring of animals for optimal reproductive performance following calving is not common on smallholder farms. Published literature on the reproductive performance of different breeds of cattle raised under smallholder farming conditions outlining details on days open, conception rates, calving intervals and their effects on milk production is limited.

The use of AI to introduce genes for improved productivity and for cross-breeding is common. However, in many instances, livestock keepers indicate a challenge with conception rates when AI is used, resulting in the farmers opting for bull services. Several smallholder farmers retain a bull within their herds. Studies on smallholder farmers in Kenya, for instance, indicate that, where dairy production using higher grade exotic animals is widespread, the use of bull services, rather than AI, is preferred due to the need for several repeat inseminations when AI is used (Baltenweck et al., 2004; Murage and Ilatsia, 2011).

3.5 Socio-economic factors

Although dairy has the potential to transform communities, realized productivity is usually affected by several social factors including ownership of land and livestock assets. In most societies, land and cattle are owned by men, while women may own smaller animals such as sheep, goats and poultry. Men and women in societies also manage different categories of cattle and are responsible for different aspects of their care. Women feed and manage calves, sick and pregnant animals; clean barns and milk cows; and play an important role in the marketing of milk, whereas men tend to be more involved in watering, managing of diseases and the sales of animals (Mekonnen et al., 2010; Herrero et al., 2013). In more recent times, however, a high proportion of men from rural areas have migrated to urban areas in search of higher wages, leaving routine livestock management practices to women, children and the elderly. Adoption of interventions aimed at intensifying livestock production such as shifting from grazing to stall feeding or keeping of potentially higher-yielding, but also more demanding, breeds, may be slow within communities as the intensification affects traditional tasks of women and would increase their workload (Roschinsky et al., 2015). The multiple

© Burleigh Dodds Science Publishing Limited, 2017. All rights reserved.

social roles of cattle in smallholder systems may also lead to compromise solutions that prevent the attainment of maximum environment and production efficiencies from dairy cattle (McDermott et al., 2010).

Productivity levels achieved in smallholder farming systems also tend to be inherently related to the poverty level of individual producers, and to the experience and exposure of the farmers to alternative technologies. Generally, the more the years of farming experience, the higher the probability of adopting new technologies.

4 Improving dairy production via breeding under smallholder systems

To achieve increased efficiency in the smallholder dairy production and to improve the competitiveness of the production systems, strategies aimed at ensuring increased productivity, rather than increased population size, need to be promoted and adopted. Smallholder farmers with limited resources are unlikely to be able to respond sustainably to increased demands for animal products without increased public investment in innovation and support platforms essential to foster technological changes required to increase productivity (Herrero et al., 2013). In this section we highlight challenges related to the adoption of breeding technologies by smallholder farmers in SSA.

4.1 Matching genotypes to production environments

One of the greatest technical challenges around optimizing utilization of breed resources in smallholder production systems in SSA is how to match livestock genotypes to production systems while taking into consideration both tangible objectives (such as increased growth rate and milk production) and less tangible objectives (such as the keeping of livestock for saving and insurance purposes, or for ceremonies and dowry).

In matching genotypes to environmental situations, issues that need to be adequately addressed include: which specific breed type and reasons for its choice, what breeding strategy would be adopted for future generations – pure breeding or cross-breeding? What level of productivity would be optimum, given the existing opportunities within the farming system targeted? (Philipsson et al., 2011). Use of genetic improvement from temperate countries may be seen an easy alternative; however, such an approach is unsuitable for two major reasons: (1) differences in the breeding objectives between importing and exporting countries and (2) presence of genotype by environment interaction (G x E) between temperate and tropical countries.

When genotypes are moved from an environment where their genetic value is known, to a similar environment elsewhere, they are likely to perform at a similar level. If, however, they are moved to a different environment, genotype by environment interactions will affect performance, and, in tropical countries, performance of high-yielding temperate breeds is often negatively affected (Ojango and Pollott, 2000; Chagunda et al., 2015). Within individual countries of SSA there exists a significant difference in the milk yields of dairy cattle, depending on the type of production system they are reared in and on the management adopted. Lowest yields are achieved under more extensive smallholder systems as illustrated for East Africa in Fig. 3. The production environment generally limits the full expression of the genetic potential of dairy cattle imported from more productive

© Burleigh Dodds Science Publishing Limited, 2017. All rights reserved.

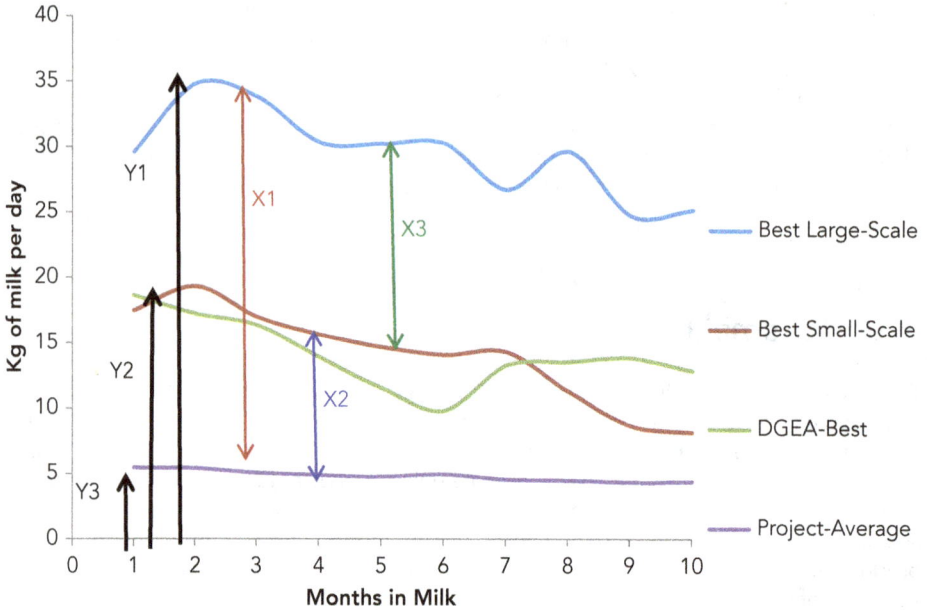

Figure 3 Lactation curves of dairy cows in Kenya demonstrating 'yield gaps' (X_1, X_2, X_3) in milk production by cattle under different farming systems (*Adapted from Dairy Genetics East Africa (DGEA), ILRI project reports*).

farming systems within the same country (Galukande et al., 2013). A more detailed analysis demonstrating potential changes in yield achievable when breeding technologies are adopted for dairy cattle raised in different developing countries, is presented by Mwacharo et al. (2009).

4.2 Cross-breeding

Cross-breeding is extensively used in an attempt to improve performance of livestock populations in developing countries (Djoko et al., 2003; Galukande et al., 2013; Leroy et al., 2015; Roschinsky et al., 2015). Generally the performance of the first cross (F_1) between the high-producing breeds from temperate environments and highly adapted breeds from tropical environments is very productive, as they are an ideal combination of both the production and adaptability. However, inter se crosses to form an F_2 are much less productive, and backcrosses to parent breeds either are poorly adapted or lack production potential (Syrstad, 1989). The lack of direction after the first cross has resulted in three main strategies being developed in the tropics: (1) continual formation of the first cross, (2) backcrossing to exotic breeds associated with improved management and (3) the formation of many informal composite populations. The use of cross-breeding for dairy production in Africa is documented in reviews by Galukande et al. (2013), FAO (2015) and Leroy et al. (2015).

© Burleigh Dodds Science Publishing Limited, 2017. All rights reserved.

4.3 Adoption and use of AI

The most widely and readily available reproductive technology used to introduce and disseminate new breeds with desired dairy characteristics in populations of SSA is AI. However, although AI has been available since the 1930s, and is relatively cheap and easy to use, it has been difficult to administer it successfully in smallholder cattle production systems mainly due to logistical and institutional challenges (Okeyo et al., 2000; Kosgey et al., 2011; Zonabend et al., 2013). Costs of semen for AI are highly variable across countries, depending on the level of development of the dairy value chain (VC), government support to the production systems, and the number of actors involved in the distribution chain (Makoni et al., 2013; Ouma et al., 2014). Additionally, the genotypes that result from the use of AI are often incompatible with the management levels and approaches in the smallholder farming systems. Small herd sizes and lack of animal pedigree and performance records result in haphazard use of semen from select sires raised under high-input environments.

In zero-grazed animals, success rates of AI are generally compromised when service providers are few. There are high chances of animals not being served, even when oestrus is correctly detected. Some countries have national semen distribution centres (Zonabend et al., 2010); however, they produce only a limited amount of semen while the rest comes from external sources. The choice of breeds for which semen is provided also tends to be driven by private agencies and perceived high levels of milk production potential by the farmers. This has resulted in widespread use of the Holstein Friesian for increasing milk production in many countries of SSA (Djoko et al., 2003; Chagunda et al., 2004; Galukande et al., 2013; Zonabend et al., 2013).

4.4 Livestock recording

Lack of comprehensive livestock data/records hugely limits planning and implementation of dairy development and breeding programmes for sustainable improvement of cattle productivity (Rege et al., 2011; Syrstad and Ruane, 1998). Livestock keepers tend to rely on their knowledge, memory and practical experiences with no written records on animal performance and management practices. Livestock records are an essential component for genetic evaluation of selection candidates, traceability of animal movements, disease control and good farm management, and contribute to securing access to markets for higher quality and geographical identifiable products (Hoffmann et al., 2011; Durr, 2012). Several constraints have been identified that limit the adoption and practice of livestock recording in developing countries (Trivedi, 2002; Banga et al., 2010; Kosgey et al., 2011; Chagunda et al., 2015), these include:

1 Inadequate and unsupportive policies and infrastructure
2 Weak or non-existent organizations and institutions to carry out and support recording systems
3 Lack of appropriate related legal frameworks resulting in inadequate and weak partnerships, networks and collaboration
4 Small and dispersed herds/flocks, leading to high transaction costs
5 Diverse and multiple stakeholders, often overlapping in their roles

© Burleigh Dodds Science Publishing Limited, 2017. All rights reserved.

6 Limited capacity and understanding of livestock recording, processing of information
   and feedback both at farmer level and at institutional level
7 Inadequate resource mobilization and allocation to support pilot activities for
   livestock recording systems

Innovative use of hand-held electronic data-capturing devices that are linked to cell
phones would greatly facilitate livestock recording.

4.5 Generating replacement animals

Breeding animals for replacement within the smallholder systems are usually few, mainly as
a result of poor reproductive management and low fertility rates of cows. A limited supply of
quality replacement heifers of the appropriate genotype for smallholder farmers constrains
scaling out dairy improvement in many countries. Lack of replacement animals also leads
to farmers retaining animals in their herds over many lactations even if their productivity is
not very high (Bebe, 2008; Banda et al., 2012; SNV, 2013). In some instances, smallholder
farmers deliberately rear and retain few young animals as potential replacement stock
in order to minimize labour and feeding costs (Bebe, 2008). There is, however, a high
demand for replacement animals within smallholder farming systems, resulting in inflated
costs of breeding animals from other sources (Leroy et al., 2015; Roschinsky et al., 2015).

5 Improving productivity in smallholder dairy systems

The 2013 GAP report (Global Harvest Initiative, 2013) notes that the adoption of advanced
agricultural technologies and better production practices are critical for realizing significant
productivity gains in both industrialized and developing countries. Intensification and
more extensive use of technologies provide an opportunity to transform dairy production
systems in Africa. In the smallholder farming systems, producers tend to intensify some
but not all aspects of their production, attracting investments in targeted technologies and
facilities for management of both the animals and the products obtained. These provide a
good opportunity for enhancing productivity of smallholder dairy systems.

5.1 Breeding and reproductive technologies

Breeding and reproductive technologies (biotechnologies) are major avenues through
which herd improvement has been achieved in developed countries. Breeding technologies
adopted over time include cross-breeding, assortative and non-assortative mating,
selection using BLUP breeding values and economic indexes. These are outlined in detail
in the second report on the state of the world animal genetic resources (FAO, 2015). Various
reproductive technologies and their impacts on breeding schemes for livestock populations
in developing countries are discussed by Van Arendonk (2011). Where smallholder farmers
can afford the inputs required, AI has positively contributed to increased productivity
(Chagunda et al., 2004; Roschinsky et al., 2015). As smallholder systems increasingly
adopt a more commercial outlook, oestrus synchronization in combination with AI using
sexed semen could be adopted. The use of sexed semen enhances a farmer's ability to
obtain replacement heifers from their own herds, and could help increase the efficiency
of producing F_1 dairy hybrids (Van Arendonk, 2011; FAO, 2015). A combination of *in*

© Burleigh Dodds Science Publishing Limited, 2017. All rights reserved.

vitro fertilization with sexed semen, followed by embryo transfer as a possible solution to increasing heifer availability, is under investigation in East Africa (Mutembei et al., 2008). The combination of using sexed semen with oestrus synchronization using hormones helped to increase the national dairy cattle herd population in Rwanda through the 'one-cow per resource-poor household' (*Girinka*) programme (Hirwa et al., 2013; Kugonza et al., 2013). Costs of reproductive technologies targeting female animals (multiple ovulation, embryo transfer, *in vitro* fertilization) are, however, significantly high (FAO, 2015) and constrain their widespread use in smallholder farming systems.

5.2 Molecular technologies

Advances in high-density single-nucleotide polymorphism (SNP) technology which enables genotyping of an individual at a low cost present an opportunity for revolutionary changes in the genetic analysis of populations and genetic improvement programmes (Schaeffer, 2006; Van Arendonk, 2011; FAO, 2015). For animals under smallholder farming systems, SNP technology offers an opportunity to reconstruct pedigrees of cross-bred animals and use genomic information in combination with phenotypic information to estimate variance components of quantitative traits. DNA samples of sires frequently used in the populations are important for facilitating pedigree reconstruction. Genomic selection offers possibilities for increasing the accuracies of breeding value estimations, lowering the rates of inbreeding, and reducing the generation intervals in dairy cattle breeding (Schaeffer, 2006).

By genotyping a large number of selection candidates and animals with performance records, one can start a selection programme using SNP-based relationships. Research is, however, needed to determine the optimal design of a breeding scheme based on SNP-based relationships from a genetic and economic point of view. Though use of molecular information and genomic selection technologies is still limited in SSA, studies are ongoing on its utilization in smallholder farming systems (Kios, van Marle-Köster and Visser, 2012; Mujibi et al., 2014; Ojango et al., 2014; Brown et al., 2016). In a simulation study that sought to optimize the design of small-sized nucleus genetic improvement dairy cattle schemes for situations where phenotype records are limited, Genomic Selection (GS)-based schemes resulted in a higher overall population mean performance with lower rates of inbreeding than progeny test schemes. However, to be optimum, the breeding scheme would rely on the annual genotyping of 5000 commercially recorded cows in the genetic evaluation (Kariuki et al., 2014).

5.3 International livestock data platforms and information and communications technologies

There is an increasing amount of information available at country level on the diversity, characteristics and use of different cattle-breed types in Africa through web-based resources including the Domestic Animal Genetic Resources Information Systems (DAGRIS, 2016) availed through ILRI, and the Domestic Animal Diversity Information System (DAD-IS, 2007) availed through the FAO. Livestock breed types that have demonstrated potential in specific production systems found on the continent need to be more extensively availed and used in targeted systems. In the East African highlands, for example, diverse mixes of cross-bred dairy cattle have resulted from over 50 years of cross-breeding efforts. There is clear evidence that some kind of cross-bred which best fits into the dairy production system has emerged (Rege et al., 2011), representing an opportunity for refined composite breed

© Burleigh Dodds Science Publishing Limited, 2017. All rights reserved.

development. A meta-analysis of information from 23 tropical countries where indigenous breeds are cross-bred with exotic *Bos taurus* breeds of dairy cattle by Galukande et al. (2013), showed that the resultant cross-breeds had higher milk yields, increased lactation lengths, shorter calving intervals and a lower age at first calving compared with the local breeds. Other potential opportunities for introducing 'new' genotypes include use of N'Dama cattle from West Africa in tsetse-infested areas of eastern Africa, the possibility for introducing Kenana and Butana of the Sudan for milk production in other low to medium potential areas of eastern Africa, and the possible use of Brazilian 'dairy' zebu breeds (for example, the Gir and Guzera) in different parts of Africa (Rege et al., 2011).

Information and communications technology tools, notably mobile telephony, can potentially transform marketing in smallholder production systems in Africa, allowing small-scale producers, processors and traders to collaboratively avail and use product and market information. The use of mobile phones within smallholder farming systems to record inputs and animal performance is currently being piloted in the East Africa region through a public and private partnership project: African Dairy Genetic Gains (http://www.ilri.org/node/40458). Training on farmer cooperation in recording is, however, required. Data collected will help guide the selection and use of sires within villages with similar environmental conditions. Recording needs to be linked to day-to-day management and used as a tool to track economics of production (Rege et al., 2011).

5.4 Feeding technologies

Development of fodder banks, improved pasture species, planted legumes and feed supplementation with crop by-products would result in better-quality diets for dairy cattle in SSA. More recently, novel livestock feeds based on crop species more indigenous to SSA are being used as an alternative sources of carbohydrates and proteins for animals (Chagunda et al., 2015). These include use of cassava roots and by-products, development of dual-purpose sorghum varieties (for grain and fodder) and use of sweet potato vines and roots as energy sources (CTA, 2015). Crop breeding for improved fodder quality can increase available feed resources and improve efficiency of feed utilization, resulting in an increase in milk production. These changes can additionally help mitigate methane emissions from smallholder dairy production systems (Herrero and Thornton, 2009).

Several manuals have been developed and updated with country- and region-specific information on feeding and management of dairy cattle in SSA, and are available to farmers through dairy development projects, or through the internet. These include manuals on feeding and managing dairy cattle in different countries of Africa (Pandey and Voskuil, 2011; Lukuyu et al., 2012; Ministry of Livestock Developement-Nairobi Kenya, 2012). Additionally, tools have been developed that provide systematic methods of assessing local feed resources available in addition to supporting the design of appropriate intervention strategies for optimal feed utilization such as ENDIISA (Mubiru et al., 2011) and FEAST (ILRI, 2014).

5.5 Delivery of breeding services

5.5.1 Breeding programmes

Community-based breeding programmes (CBBP) offer opportunities for effectively involving local farmers. Examples of CBBP in use in different countries for different species of livestock are presented by Mueller et al. (2015). Improvement in dairy productivity

© Burleigh Dodds Science Publishing Limited, 2017. All rights reserved.

under smallholder systems can be created in locally available genetic resources using cross-breeding as a first stepping stone, and AI to disseminate improved genetic material to other farmers. Developing different breeds specialized for the different environments under which smallholder dairy production is practised would help maintain genetic diversity while improving productivity.

Nucleus-based models can also be adapted in which the available adapted dairy genetic resources are concentrated in a few smallholder herds for selective breeding. Rege et al. (2011) proposed a model consisting of 'cattle breeding units' run by individuals or groups specializing in the production of desired genotypes (e.g. purebreds or F_1 crosses) – akin to the day-old-chick model in poultry – for smallholder dairy production systems. Kahi et al. (2004) illustrated how a well-organized nucleus could be established as a private enterprise owned by farmer cooperatives utilizing the smallholder farmers as the commercial population. Kahi and Nitter (2004) proposed a two-tier nucleus breeding scheme using young sires for dairy production systems in developing countries. Smallholder farmers could serve as the commercial population in which improvement achieved under nucleus conditions on larger-scale, more commercial enterprises would be utilized. The nucleus could also be in the form of a private enterprise owned by farmer cooperatives or individuals. Options for smallholder farmers to remain competitive would include their linking up to grow in scale through cooperative arrangements, or their becoming contract farmers to larger operations (Delgado et al., 2008). For nucleus breeding programmes to be successful, business plans need to be developed to provide direction on the generation of income in the long run through the marketing of breeding stock and/or culled animals (Mwacharo and Drucker, 2005).

5.5.2 Production and access to improved sires

Use of locally produced semen from young bulls, sons of sires tested within a country has been shown to be more cost-effective than using imported semen from sires raised in more temperate environments (Okeno, Kosgey and Kahi, 2010). Innovative contractual arrangements would, however, be required to link the multiple smallholder livestock producers. Smallholder farmers within a village using collectively determined cost-sharing methods could own and manage select bulls for use in breeding within the village. Such a scheme could be adapted to include rotation or swapping of bulls across villages in order to avoid potential inbreeding within the smallholder herds.

An increasing interest by private sector actors, willing to invest in the development, adoption and dissemination of targeted reproductive technologies for increasing profits from smallholder dairy production also provides expanded options for choice of breeds and sires to use in AI programmes. Using business model options in which farmer advisory services are combined with AI services for a large number of smallholder farmers, economies of scale are achievable for dissemination of breeding sires across populations.

5.5.3 Models for heifer production

Options for availing quality heifers of the right genotype, at a price that smallholder farmers or new entrants wishing to establish dairy farming at small or even medium scale, can afford, notably in eastern Africa include:

- *Breed Heifers from within the herd*: While this is the most logical approach to producing replacement heifers, it is not practical for smallholders because the herd

© Burleigh Dodds Science Publishing Limited, 2017. All rights reserved.

sizes are too small to guarantee that heifer calves will be born and successfully raised for use as replacement animals when they are needed.

- *Contract mating for cross-bred heifer production*: This approach emerged from an innovation platform (IP) in East Africa and is dubbed 'wombs for hire' (AgriBiz Consult Company Kenya, pers. comm.). The approach involves mating of quality beef cows (e.g. from ranches or improved pastoral systems) with dairy bulls (or AI semen) in a contractual arrangement with the aim of producing cross-bred heifer calves of specific breed combinations. The contract clearly stipulates the obligations of the parties (ranch and the *heifer business* entity) in regard to the resulting calves – both heifers and bulls. The link with larger-scale farming systems ensures a reliable supply of a targeted number of improved animals which can be availed on given orders. Combined with sexed semen, this approach has potential to address the major heifer supply deficit that characterizes smallholder systems in eastern Africa.

- *Calf nursery model*: This model seeks to address the inability of smallholder farmers to rear good calves because of limited resources in terms of space, time, and milk (which is valued for sale) or milk-replacers (which these farmers cannot afford). Under the calf nursery model, the operator (e.g. a medium-scale farmer) goes into an arrangement with smallholder farms so that heifer calves born on smallholder farms are taken away soon after being reared on colostrum and are reared under the best possible conditions to develop them into high-quality heifers.

- *Online animal trading platforms*: There is emergence of online platforms and mobile-phone-based applications designed to connect buyers to sellers and service providers to livestock keepers. In the livestock sector, commodity specific platforms include I-COW (http://www.icow.co.ke/) and CowSoko (http://www.cowsoko.com/). Using simple menu systems, farmers can access improved heifers from a broader population, and are in a position to better negotiate for prices of animals with knowledge of their availability.

- *Pass on the gift model*: This approach of transforming recipients into donors was spearheaded in the dairy sector by Heifer International (Njwe et al., 2001). The approach mandates that each family that receives a donation in the form of a heifer undertakes to pass on to a neighbour (member of a group) the first heifer calf produced by the heifer received. This approach which proves that small actions can achieve big results, at a minimum, has potential to double the impact of the original gift, transforming a once-impoverished family/community into full participants who improve and strengthen the bonds within their broader communities.

5.6 Support groups for smallholder dairy production

In many countries, national governments have livestock ministries to support dairy production (FAO, 2007; Zonabend et al., 2013; Leroy et al., 2015). However, resources to support extension services and availing inputs for different farming systems are limited. Governments thus generally provide a policy framework for operation. Producer organizations play an important role for smallholder dairy farmers. Through coming together in organized groups, the smallholder farmers can seek services and knowledge support to help in their operations. Different types of organized groups exist. Initial groups are generally formed around the marketing of products; however, more recently, groups have been formed to seek information and lobby for services and inputs to support dairy

© Burleigh Dodds Science Publishing Limited, 2017. All rights reserved.

production. Cooperation between farmers is crucial, especially in light of the change in demand patterns for livestock products, and of the growing regional economies. Examples of cooperating farmer groups are as follows.

5.6.1 Dairy cooperatives

Cooperatives in the dairy sector of SSA are membership organizations, which are generally organized around the marketing of milk by concentrations of farmers. Cooperatives provide focal points to provide services to farmers as well as promoting organized collection, handling and sale of milk to consumers. In many countries of SSA, notably countries of Eastern and Southern Africa, cooperatives have served to catalyse increased milk production from smallholder farms through access to broader markets for the products. They enable improving competition, product quality and open market economics in the dairy sectors. Cooperatives also provide a safety outlet for farmers to sell their milk through bulking and marketing of the products. Many cooperatives, however, need strengthening in terms of management, governance, negotiation capacities and organization.

5.6.2 Dairy hubs

Dairy hubs (also referred to as milk bulking groups (Tebug et al., 2012)) are a growing concept used in areas with a high concentration of smallholder dairy farmers to stimulate productivity and reduce dependency. The hub concept is based on linking farmers in a specific area – covering a certain number of villages, smallholder farmers and cows – to a dedicated dairy processor. The processor sets up milk collection stations (hubs) with cooling tanks for bulking and controlling quality of milk collected from farmers. The hubs additionally serve as business centres for dairy farmers to access a host of services through a check-off system. Under the check-off system, farmers deliver milk to their dairy hub and secure inputs (feeds, drugs) and services (credit, AI, veterinary) on account of their deliveries. The cost of the inputs and services is then deducted from their milk income at the end of the month when collecting payment (ILRI, 2008; SNV, 2013).

The hubs minimize the cost of collecting milk from small, scattered producers by the major processing firms who pay a premium price for bulked and chilled milk. Through the hubs, processors are able to tap into a reliable supply of locally produced, high-quality milk and gain better control over the supply chain. At the same time, public access to safe and affordable milk is increased. The dairy hub model has been successfully introduced and promoted in eastern Africa through development projects such as the East Africa Dairy Development (EADD) programme in Kenya, Uganda, Rwanda and Tanzania (ILRI, 2008), and in Malawi.

5.7 Ensuring VCs work for smallholders

Rege et al. (2011) make a compelling case that institutional arrangements and enabling policies are critical for the success in identifying and applying appropriate genetic technologies, improving access to input services and facilitating access to markets in order to translate productivity gains into incomes for smallholders. The main actors involved in smallholder dairy VC include individual producers (farmers) or producer groups, associations or cooperatives, policy-makers, and implementers and regulators;

© Burleigh Dodds Science Publishing Limited, 2017. All rights reserved.

input suppliers; other VC service providers; market agents; research and development organizations; and development partners as well as non-governmental organizations supporting smallholders. Many challenges in the smallholder dairy VC can only be addressed if the VC actors worked together. For example, while smallholder systems are generally associated with low production costs and potential for high profit margins, limited access to quality, affordable and dependable inputs, services and high value and reliable output markets, limit the potential of this important sub-sector. Indeed, the viability and profitability of milk production by smallholders depends not only on production costs and what individual producers can do on their own, but more critically on the efficiency of the overall dairy VC in which the producers are involved (McDermott et al., 2010). Organized small-scale dairy systems can compete successfully with large-scale dairies and, in some aspects, can outdo the individual large-scale operators if a functioning VC is developed.

A functioning VC is a market-focused collaboration among the different stakeholders involved in the VC. IPs (Hall and Mbabu, 2012; Hounkonnou et al., 2012) can be effective as a means of organizing functioning VCs. An IP facilitates the bundling of complementary skills and competencies that the VC actors bring – and that are linked to their core businesses, expertise and experiences as individuals, teams or organizations – and allows for working on institutional change at different levels in the system. IPs ensure that different interests are taken into account, and various groups contribute to finding solutions.

An IP formed to enhance VC functionality takes the form of a 'partnership' whose basic construct is defined by direct and honest engagement of VC actors to facilitate ongoing sharing, problem identification and co-creation of solutions to common challenges whose impacts are felt by multiple VC actors. In the process, individual entities (e.g. producers) also learn how to improve their own businesses and at the same time expand their networks. The success of the VC will be principally defined by core VC actors who stand to gain if the VC is flourishing and to lose if it is dysfunctional. The IP provides a formal forum through which this recognition plays out repeatedly.

6 Key organizations supporting smallholders

Continental organizations in place to support the development of the livestock sector in Africa, and which provide guidance and mobilize resources for the adoption and use of various technological applications for improved productivity include The African Union–Inter-African Bureau for Animal Resources (AU-IBAR, http://www.au-ibar.org/) and the Forum for Agricultural Research in Africa (FARA, http://faraafrica.org/). These organizations work together with sub-regional organizations including the Association for Strengthening Agricultural Research in East and Central Africa (ASARECA, http://harvestchoice.org/regions/ASARECA), the West and Central Africa Council for Agricultural Research and Development (CORAF/WECARD, http://www.coraf.org/en/) and the Centre for Coordination of Agricultural Research and development for Southern Africa (CARDESA, http://www.ccardesa.org/). These organizations serve to mobilize stakeholders around a portfolio of programmes and projects seeking to address specific challenges or harness opportunities.

© Burleigh Dodds Science Publishing Limited, 2017. All rights reserved.

7 Future trends

Long-term sustainability of improvement of dairy cattle productivity under smallholder farming systems of SSA still remains a challenge. The smallholder farmers tend to lapse on the implementation of the management regimes instituted through projects once the projects come to an end (Bebe, 2008; Roschinsky et al., 2015). The small scale of milk production and marketed output implicit in smallholder farming systems can result in low bargaining power and a limited ability to capture economies of scale in marketing. Inadvertently, attention needs to be paid both to the production side and to the complexity in the VC including simple methods of value addition and continued access to services and markets for products (Staal, Pratt and Jabbar, 2008). External factors are, however, catalysing change in existing smallholder production systems of SSA:

- *Population pressure*: As the human population increases, the demand for products increases, thus motivating livestock keepers to increase their productivity. This drives uptake of productivity-enhancing interventions but also affects the available land and water resources needed to support commercial dairying.
- *Public infrastructure development*: Road networks, electricity and fuel availability, and communication networks, all these influence every component of the dairy VC, and particularly stimulate access to inputs, markets for livestock products and access to administrative and technical services.

Intervention options to enable sustainable change in smallholder dairy production systems can be categorized into five main clusters outlined below and illustrated in Fig. 4.

1 Develop and adopt *transformative approaches* to generate evidence on the magnitude and impacts of smallholder dairy production on national food security and regional stability. Such evidence should inform policy development and drive investments in the smallholder farming sector. Pilot farmers using an agricultural innovation systems approach in the smallholder dairy production VC could serve to catalyse adoption of new approaches.

2 *Empower* smallholder livestock producers to enable them improve the efficiency of their dairy production through seeking and adopting demand-driven inputs and services for the sector. With less than 50% of the milk produced by smallholder farmers in Africa being sold through formal (processed) channels, the safety of dairy products is becoming an increasing concern. Urbanization and growth in incomes has resulted in increased consumer demand for food safety, and formal marketing of dairy products is important in the livestock food chain. Smallholder farmers need to adapt to the changes in the VC in order to improve their credibility (Delgado et al., 2008).

3 Support *inclusive and collective actions* by organized groups operating within the smallholder dairy VC. This could be in the form of frameworks that support and generate confidence in private–public partnerships; advocacy at national and regional levels for raising profiles on opportunities and causes of concern for smallholder dairy farmers; engagement of a new and younger generation of farmers who are willing and able to collate, share and use information via information and communications technology platforms.

© Burleigh Dodds Science Publishing Limited, 2017. All rights reserved.

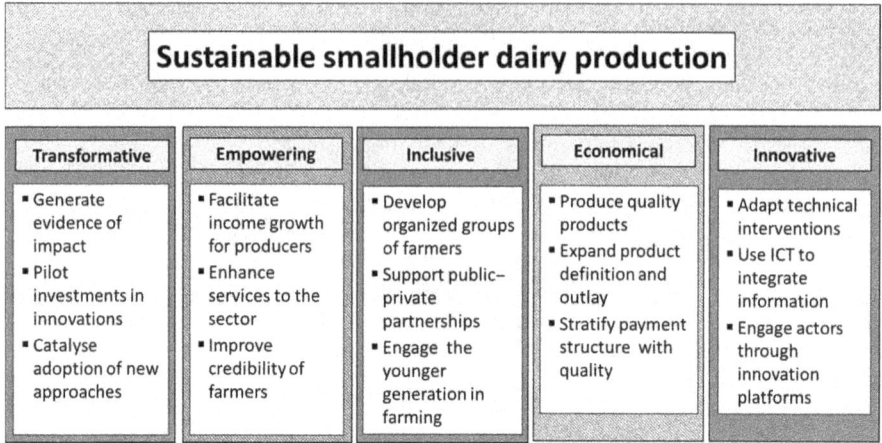

Figure 4 Interventions for sustainable smallholder dairy production.

4   *Enhance markets* for quality dairy products and develop stratified markets with system specific improved breeding animals for optimal productivity. Milk component traits rarely emphasized in the production systems need to be valued and milk pricing-based quality rather than on volume. Markets should have graded product levels and requisite payment for producers able to avail products of specific quality standards in the marketplace.

5   *Innovatively adapt* technological interventions in a cost-effective manner to suit smallholder production systems. One strength of smallholder dairy producers that is often overlooked as an asset in terms of investment by the private sector is the large number of producers. Used strategically, these producers could provide a significantly large population of test animals with wide variation that could serve to prove robustness of different technologies. Important are the IPs that bring together different actors in the dairy production sector to raise issues and to propose solutions to challenges encountered.

8 Where to look for further information

An increasing amount of information on livestock production systems in developing countries is being generated and published in peer-reviewed open source journals, including the *Animal Genetic Resources* (FAO), *Tropical Animal Production and Health* and *Livestock Research for Rural Development*. There are also several international agricultural research institutes that carry out research and document information on livestock farming systems in developing countries, including the Consultative Group on International Agricultural Research (CGIAR) centres (International Livestock Research Institute (ILRI, http://www.ilri.org), International Center for Agricultural Research in the Dry Areas (ICARDA), the International Center for Tropical Agriculture (CIAT)), ICIPE, and the Food and Agriculture Organization (FAO, http://www.fao.org). The individual countries of SSA

© Burleigh Dodds Science Publishing Limited, 2017. All rights reserved.

through national universities and national agricultural research centres also implement research on, and document information unique to, their existing cattle populations.

Rapid on-farm surveys implemented by a range of research groups address critical information gaps on production systems, population statistics of breeds, physical descriptive characteristics and prevailing performance levels of animals reared under smallholder farming systems (ILRI rapid surveys, FAO guidelines, AGTR Case studies) (Nkya et al., 2007; Mekonnen, Dehninet and Kelay, 2010; Tebug et al., 2012).

9 Acknowledgements

The authors acknowledge support for this study through the International Livestock Research Institute (ILRI) and the African Union–Inter-African Bureau for Animal Resources (AU-IBAR).

10 References

Baltenweck, I., Ouma, R., Anunda, F., Okeyo, M. and Romney, D. (2004), 'Artificial or natural insemination: the demand for breeding services by smallholders', in *Proceedings of 9th KARI Biennial Scientific Conference and Research week*. 8–12 November 2004, Nairobi, Kenya.

Banda, L. J., Kamwanja, L. A., Chagunda, M. G. G., Ashworth, C. J. and Roberts, D. J. (2012), 'Status of dairy cow management and fertility in smallholder farms in Malawi', *Tropical Animal Health and Production*, 44(4), pp. 715–27.

Banga, C. B., Besbes, B., Balvay, B., Chazo, L., Jamaa, O. M., Rozstalynyy, A., Rovere, G., Toto, A. and Trivedi, K. R. (2010), 'Current situation of animal identification and recording systems in developing countries and countries with economies in transition', in E. Skujina, E. Galvanoska, O. Leray and C. Mosconi (Eds), *ICAR Technical Series No 14. Farm Animal Breeding, Identification, Production Recording and Management*. ICAR, Rome, Italy.

Bebe, B. O. (2008), 'Assessing potential for producing dairy replacements under increasing intensification of smallholder dairy systems in the Kenya highlands', *Livestock Research for Rural Development*. Retrieved 22 April 2016, p. 20.

Brown, A., Ojango, J., Gibson, J., Coffey, M., Mwai, O. and Mrode, R. (2016), 'Genomic selection in a crossbred cattle population using data from the Dairy Genetics Project for East Africa', *Journal Dairy Science*, 99, pp. 1–5.

Chagunda, M.G.G., Bruns, E.W., King, J.M. and Wollny, C.B.A. (2004), 'Evaluation of the breeding strategy for milk yield of Holstein Friesian cows on large-scale dairy farms in Malawi', *Journal of Agricultural Science*, 142, pp. 595–601.

Chagunda, M.G.G., Bruns, E., Wollny, C.B.A. and King, H.M. (2004), 'Effect of milk yield-based selection on some reproductive traits of Holstein Fresian cows on large scale dairy farms in Malawi', *Livestock Research for Rural Development*, 16(7), pp. 20–32. Available at: http://www.cipav.org.co/lrrd/lrrd16/7/chag16047.htm.

Chagunda, M.G.G., Mwangwela, A., Mumba, C., Dos Anjos, F., Kawonga, B.S., Hopkins, R. and Chiwona-Kartun, L. (2015), 'Assessing and managing intensification in smallholder dairy systems for food and nutrition security in Sub-Saharan Africa', *Regional Environmental Change*. Springer Berlin, Heidelberg.

CTA (2015), *Fodder and Forage Solutions – Feeding Africa's Livestock. Technical Centre for Agricultural and Rural Cooperation ACP-EU*, Wageningen, The Netherlands. Available at: http://archive.spore.cta.int/en/component/content/article?id=11543:feeding-africa-s-livestock.

© Burleigh Dodds Science Publishing Limited, 2017. All rights reserved.

DAD-IS (2007), *Domestic Animal Diversity Information System*, FAO. Available at: dad.fao.org/cgi bin/EfabisWeb.cgi?

DAGRIS (2016), *Domestic Animal Genetic Resrouces Information System*, ILRI. Available at: http://dagris.ilri.cgiar.org/dagrisv3/default.asp.

Devendra, C. (2001a), 'Small ruminants: imperatives for productivity enhancement improved livelihoods and rural growth – a review', *Asian-Australasian Journal of Animal Science*, 14(10), pp. 1483–96.

Devendra, C. (2001b), 'Smallholder dairy production systems in developing countries: characteristics, potential and opportunities for improvement – a review', *Asian-Australasian Journal of Animal Sciences*, 14, pp. 104–13.

Djoko, T. D., Mbah, D. A., Mbanya, J. N., Kamga, R., Awah, N. R. and Bopelet, M. (2003), 'Crossbreeding cattle for milk production in the tropics: effects of genetic and environmental factors on the performance of improved genotypes on the cameroon western high plateau', *Revue d Elevage et de Medecine Veterinaire des Pays Tropicaux*, 56(1–2), pp. 63–72.

Durr, J. (2012), 'Animal identification, recording and evaluation – Interbull's perspectives', http://www.icar.org/Documents/Santiago%202011/Papers/Durr.pdf.

FAO (2007), *The State of the World's Animal Genetic Resources for Food and Agriculture*. Edited by R. Burbeck and D. Pilling, Food and Agriculture Organization of the United Nations, Rome, Italy.

FAO (2015), *The Second Report on the State of the World's Animal Genetic Resources for Food and Agriculture*. Edited by B. D. Scherf and D. Pilling, FAO Commissnion on Genetic Resources for Food and Agriculture Assessments, Rome, Italy. Available at: http://www.fao.org/3/a-i4787e/index.html.

FAO (2016), *Dairy Producion and Products*. Available at: http://www.fao.org/agriculture/dairy-gateway/milk-production/production-systems/en/#.Vs6qguZL8ic (Accessed 12 July 2016).

FAOSTAT (2014), 'FAO Statistical Database', FAO, Rome, Italy. Available at: http://faostat.fao.org.

FAOSTAT (2016), 'FAOSTAT Database', FAO, Rome, Italy. Available at: http://apps.fao.org/ (Accessed 29 Septmeber 2016).

Forbes, B. and Kepe, T. (2014), 'Smallholder farmers' attitudes toward the provision of drinking water for dairy cows in Kagera, Tanzania', *Tropical Animal Health and Production*, 47(2), pp. 415–21.

Galukande, E., Mulindwa, H., Wurzinger, M., Roschinsky, R., Mwai, A. O. and Sölkner, J. (2013), 'Cross-breeding cattle for milk production in the tropics: achievements, challenges and opportunities', *Animal Genetic Resources/Ressources génétiques animales/Recursos genéticos animales*, 52, pp. 111–25.

Global Harvest Initiative (2013), *The 2013 Global Agricultural Productivity (GAP) Report*. Washington, DC.

Hall, A. and Mbabu, A. N. (2012), *Capacity Building For Agricultural Research For Development: Lessons From Practice in Papua New Guinea*. Edited by A. Hall and A. N. Mbabu. Maastricht, UNU-MERIT.

Herrero, M., Grace, D., Njuki, J., Johnson, N., Enahoro, D., Silvestri, S. and Rufino, M. C. (2013), 'The roles of livestock in developing countries', *Animal*, 7, pp. 3–18.

Herrero, M. and Thornton, P. K. (2009), *Mitigating Greenhouse Gas Emissions from Livestock Systems. Agriculture and Climate Change: An Agenda for Negotiation in Copenhagen*. IFPRI 2020 Vision for Food, Agriculture, and the Environment, Focus 16. Washington, International Food Policy Research In. Report.

Herrero, M., Thornton, P. K., Notenbaert, A., Msangi, S., Wood, S., Kruska, R., Dixon, J., Bossio, D., Van de Steeg, J. A., Freeman, H. A. and Parthasarathy Rao, P. (2009), *Drivers of Change in Crop–Livestock Systems and their Potential Impacts on Agro-Ecosystems Services and Human Well-Being to 2030, Nairobi*. Report. CGIAR Systems and their programme, ILRI, Nairobi, Kenya.

Herrero, M., Thornton, P. K., Notenbaert, A. M., Wood, S., Msangi, S., Freeman, H. A., Bossio, D., Dixon, J., Peters, M., van de Steeg, J., Lynam, J., Parthasarathy Rao, P., Macmillan, S., Gerald, B., McDermott, J., Seré, C. and Rosegrant, M. (2010), 'Smart investments in sustainable food production: revisiting mixed crop-livestock systems', *Science*, 327, pp. 822–5.

© Burleigh Dodds Science Publishing Limited, 2017. All rights reserved.

Hirwa, C. D., Kugonza, D. R., Manzi, M., Murekezi, T., Nyabinwa, P., Semahoro, F., Rwemarika, D., Nshimiyimana, A. M. and Gahakwa, D. (2013), *Efficacy of Progesterone Releasing Intra-Vaginal Device (Prid) In Pure Ankole Cattle And Crossbreds With Three Exotic Breeds.* Centre for Agriculture and Bioscience International (CABI), Oxfordshire, UK.

Hoffmann, I., Besbes, B. and Battaglia, D. (2011), 'Capacity building in support of animal identification for recording and traceability: FAO's multipurpose and global approach', Animal Production and Health Division, Food and Agriculture Organization of the United Nations (FAO), Rome, Italy.

Hounkonnou, D., Kossou, D., Kuyper, T. W., Leeuwis, C., Nederlof, E. S., Röling, N., Sakyi-Dawson, O., Traoré, M. and Van Huis, A. (2012), 'An innovation systems approach to institutional change: smallholder development in West Africa', *Agricultural Systems*, 108, pp. 74–83.

ILRI (2008), *The East Africa Dairy Development (EADD) Project.* Available at: http://www.ilri.org/ EADD (Accessed 1 November 2015).

ILRI (2014), 'FEAST: Feed Assessment Tool', International Livestock Research Institute, Nairobi, Kenya. Available at: http://hdl.handle.net/10568/16539.

ILRI (2016), *Animal BioSciences.* Available at: https://www.ilri.org/taxonomy/term/26, https://www.ilri. org/taxonomy/term/28, https://www.ilri.org/taxonomy/term/41.

Kahi, A. K. and Nitter, G. (2004), 'Developing breeding schemes for pasture based dairy production systems in Kenya. I. Derivation of economic values using profit functions', *Livestock Production Science*, 88, pp. 161–77.

Kahi, A. K., Nitter, G. and Gall, C. F. (2004), 'Developing of breeding schemes for pasture based dairy production systems in Kenya. II. Evaluation of alternative objectives and schemes using a two- tier open nucleus and young bull system', *Livestock Production Science*, 88, pp. 179–92.

Kariuki, C. M., Komen, H., Kahi, A. K. and van Arendonk, J. A. M. (2014), 'Optimizing the design of small-sized nucleus breeding programs for dairy cattle with minimal performance recording', *Journal of Dairy Science*, 97, pp. 7963–74.

Kios, D., van Marle-Köster, E. and Visser, C. (2012), 'Application of DNA markers in parentage verification of Boran cattle in Kenya', *Tropical Animal Health and Production*, 44(3), pp. 471–6.

Kosgey, I. S., Mbuku, S. M., Okeyo, A. M., Amimo, J., Philipsson, J. and Ojango, J. M. K. (2011), 'Institutional and organizational frameworks for dairy and beef cattle recording in Kenya: a review and opportunities for improvement', *Animal Genetic Resources. Food and Agriculture Organization of the United Nations*, 48, pp. 1–11.

Kugonza, D. R., Kayitesi, A., Semahoro, F., Ingabire, D., Manzi, M., Hirwa, C. A. and Gahakwa, D. (2013), 'Factors affecting suitability of surrogate dams for embryo transfer in cattle', *Journal of Animal Science Advances*, 3(4), pp. 203–10.

Leroy, G., Baumung, R., Boettcher, P., Scherf, B. and Hoffmann, I. (2015), 'Review: sustainability of crossbreeding in developing countries; definitely not like crossing a meadow', *Animal* (November), pp. 1–12.

Lukuyu, B., Gachuiri, C. K., Lukuyu, M. N., Lusweti, C., Mwendia, S., Gachuiri, C. and Lukuyu, M. (2012), 'Feeding dairy cattle in East Africa', p. 2. Available at: https://cgspace.cgiar.org/ bitstream/handle/.../EADDDairyManual.pdf.

Makoni, N., Mwai, R., Redda, T., van der Zijpp, A. and van der Lee, J. (2013), 'White gold: opportunities for dairy sector development collaboration in east africa', Centre for Development Innovation (February), p. 203.

Marshall, K., Tebug, S., Juga, J., Tapio, M. and Missohou, A. 2016. 'Better dairy cattle breeds and better management can improve the livelihoods of the rural poor in Senegal', *ILRI Research Brief 65*. ILRI, Nairobi, Kenya. Available at http://hdl.handle.net/10568/72865

Mburu, L. M., Wakhungu, J. W. and Kang'ethe, W. G. (2007), 'Characterization of smallholder dairy production systems for livestock improvement in Kenya highlands', *Livestock Research for Rural Development*, 19, p. 110. Available at: http://www.lrrd.org/lrrd19/8/mbur19110.htm.

McDermott, J. J., Staal, S. J., Freeman, H. A., Herrero, M. and Van de Steeg, J. A. (2010), 'Sustaining intensification of smallholder livestock systems in the tropics', *Livestock Science*, 130(1–3), pp. 95–109.

© Burleigh Dodds Science Publishing Limited, 2017. All rights reserved.

Mekonnen, H., Dehninet, G. and Kelay, B. (2010), 'Dairy technology adoption in smallholder farms in "Dejen" district, Ethiopia', *Tropical Animal Health and Production*, 42(2), pp. 209–16.

Menjo, D. K., Bebe, B. O., Okeyo, A. M. and Ojango, J. M. K. (2009), 'Survival of Holstein-Friesian heifers on commercial dairy farms in Kenya', *Applied Animal Husbandry & Rural Development*, 2, pp. 14–17.

Ministry of Livestock Developement-Nairobi Kenya (2012), *Dairy Farmers Training Manual 1*. Nairobi, Kenya.

Moll, H. A. J., Staal, S. J. and Ibrahim, M. N. M. (2007), 'Smallholder dairy production and markets: a comparison of production systems in Zambia, Kenya and Sri Lanka', *Agricultural Systems*, 94(2), pp. 593–603.

Msangi, B. S. J., Bryant, M. J., Nkya, R. and Thorne, P. J. (2004), 'The effects of a short-term increase in supplementation on the reproduction performance in lactating crossbred dairy cows', *Tropical Animal Health and Production*, 36(8), pp. 775–87.

Mubiru, S. L., Wakholi, P., Nakiganda, A., Sempebwa, H. N., Namagembe, A., Semakula, J., Lule, A. and Kazibwe, P. (2011), 'Development of Endiisa decision support tool for improved feeding of dairy cattle in Uganda', in *CTA and FARA. Agricultural Innovations for Sustainable Development. Contributions from the Finalists of the 2009/2010 Africa-wide Women and Young Professionals in Science Competitions*. Accra, Ghana, pp. 45–50.

Mueller, J. P., Rischkowsky, B., Haile, a., Philipsson, J., Mwai, O., Besbes, B., Valle Zárate, A., Tibbo, M., Mirkena, T., Duguma, G., Sölkner, J. and Wurzinger, M. (2015), 'Community-based livestock breeding programmes: essentials and examples', *Journal of Animal Breeding and Genetics*, 132(2), pp. 155–68.

Mujibi, F. D. N., Ojango, J., Rao, J. E. O., Karanka, T., Kihara, A., Marete, A., Baltenweck, I., Poole, J., Rege, J. E. O., Gondro, C., Weerasinghe, W. M. S. P., Gibson, J. P. and Okeyo, A. M. (2014), 'Use of high density snp genotypes to determine the breed composition of cross bred dairy cattle in smallholder farms: assessment of reproductive and health performance', in *Proceedings of 10th World Congress on Genetics Applied to Livestock Production*, pp. 4–6.

Murage, A. W. and Ilatsia, E. D. (2011), 'Factors that determine use of breeding services by smallholder dairy farmers in Central Kenya', *Tropical Animal Health and Production*, 43(1), pp. 199–207.

Mutembei, H. M., Muasa, B., Origa, R., Jimbo, S., Ojango, J. M. K., Tsuma, V. T., Mutiga, E. R. and Okeyo, A. M. (2008), 'Delivery of appropriate cattle genotypes to Eastern African smallholder farmers through in-vitro embryo production technologies- the technical procedures, prospects and challenges', in *Pan AFRICAN Chemistry Network-Biodiversity Conference*. 10–12 September 2008, Chiromo, Nairobi, Kenya.

Mwacharo, J. M. and Drucker, A. G. (2005), 'Production objectives and management strategies of livestock keepers in South-East Kenya: implications for a breeding programme', *Tropical Animal Health and Production*, 37, pp. 635–52.

Mwacharo, J. M., Ojango, J. M. K., Baltenweck, I., Wright, I., Staal, S., Rege, J. E. O. and Okeyo, A. M. (2009), *Livestock Productivity Constraints and Opportunities for Investment in Science and Technology*, BMGF-ILRI Project on Livestock Knowledge Generation. Report. Nairobi, Kenya, ILRI.

Njwe, R. M., Kwinji, L. N., Gabche, A. L. and Tambi, E. N. (2001), 'Contributions of Heifer Project International (HPI) to small-scale dairy development in Cameroon', in D. Rangnekar and W. Thorpe (Eds), *South-South Workshop held at NDDB*. 13–16 March 2001, NDDB (National Dairy Development Board), Anand, India and ILRI (International Livestock Research Institute), pp. 414–30.

Nkya, R., Kessy, B. M., Lyimo, Z. C., Msangi, B. S. J., Turuka, F. and Mtenga, K. (2007), 'Constraints on smallholder market oriented dairy systems in the north eastern coastal region of Tanzania', *Tropical Animal Health and Production*, 39(8), pp. 627–36.

Ojango, J. M. K., Marete, A., Mujibi, D., Rao, J., Pool, J., Rege, J. E. O., Gondro, C., Weerasinghe, W. M. S. P., Gibson, J. P. and Okeyo, A. M. (2014), 'A novel use of high density SNP assays to optimize choice of different crossbred dairy cattle genotypes in small-holder systems in East

© Burleigh Dodds Science Publishing Limited, 2017. All rights reserved.

Africa', *Proceedings of 10th World Congress of Genetics Applied to Livestock Production*, pp. 2–4.

Ojango, J. M. K. and Pollott, G. (2000), 'Effects of Genotype by Environment Interactions on milk production between Holstein-Friesians in Kenya and the United Kingdom', in *3rd All Africa Conference on Animal Agriculture and 11th Conference of the Egyptian Society of Animal Production*, pp. 657–61.

Okeno, T. O., Kosgey, I. S. and Kahi, A. K. (2010), 'Genetic evaluation of breeding strategies for improvement of dairy cattle in Kenya', *Tropical Animal Health and Production*, 42(6), pp. 1073–9.

Okeyo, A. M., Kajume, J. K., Mosi, R. O., Okila, E. V. A., Gathuma, J. M., Kiere, S. M. N., Agumbah, G., Kuria, J. N. and Chema, S. (2000), 'Artificial Insemination a bio-technological tool for genetic improvement of Kenyan dairy cattle herds: historical perspective, Current status, Challenges and Way forward in the next Millenium. A Kenya Country Paper', in *Symposium on Dairy Cattle Breeding in East Africa: Sustainable Artificial Insemination Service*. Kenya Agricultural Research Institute (KARI), Headquarters, Kaptagat Road, Nairobi, 20–21 March 2000.

Ouma, R., Jakinda, D., Magati, P. and Rege, J. E. O. (2014), *Benchmarking the Kenyan Artificial Insemination Service Sub-Industry. A Study for the Kenya Markets Trust and the Competition Authority of Kenya*. Nairobi, Kenya.

Pandey, G. S. and Voskuil, G. C. J. (2011), *Manual on Improved Feeding of Dairy Cattle By*, Golden Valley Agricultural Research Trust, Lusaka, Zambia, p. 52.

Peeler, E. J. and Omore, A. O. (1997), *Manual of Livestock Production Systems in Kenya KARI/DFID National Agricultural Research Project II*, Nairobi, Kenya.

Philipsson, J., Rege, J. E. O., Zonabend, E. and Okeyo, A. M. (2011), 'Sustainable breeding programmes for tropical farming systems', in J. M. Ojango, B. Malmfors and A. M. Okeyo (Eds), *Animal Genetics Training Resource*, version 3. International Livestock Research Institute, Nairobi, Kenya and Swedish University of Agricultural Sciences, Uppsala, Sweden.

Rege, J. E. O., Kahi, A. K., Okomo-Adhiambo, M. O., Mwacharo, J. and Hanotte, O. (2001), *Zebu cattle of Kenya: Uses, Performance, Farmer Preference, Measures of Genetic Diversity and Options for Improved Use*, Animal Genetic Resources Research, ILRI, Nairobi, Kenya.

Rege, J. E. O., Marshall, K., Notenbaert, A., Ojango, J. M. K. and Okeyo, A. M. (2011), 'Pro-poor animal improvement and breeding – what can science do?', *Livestock Science*, 136, pp. 15–28.

Rege, J. E. O. and Tawah, C. L. (1999), 'The state of African cattle genetic resources II. Geographical distribution, characteristics and uses of present-day breeds and strains', *Animal Genetic Resources Information*, 26, pp. 1–25. Available at: http://agtr.ilri.cgiar.org/documents/Library/docs/agri26_99.pdf.

Reynolds, L., Metz, T. and Kiptarus, J. (1996), 'Smallholder dairy production in Kenya', *World Animal Review (FAO)*, 87–1996/2. Available at: http://www.fao.org/docrep/W2650T/W2650t07.htm#P0_0.

Robinson, T. P., Thornton, P. K., Franceschini, G., Kruska, R. L., Chiozza, F., Notenbaert, A., Cecchi, G., Herrero, M., Epprecht, M., Fritz, S., You, L., Conchedda, G. and See, L. (2011), *Global Livestock Production Systems*. Food and Agriculture Organization of the United Nations (FAO) and International Livestock Research Institute (ILRI).

Roschinsky, R., Kluszczynska, M., Sölkner, J., Puskur, R. and Wurzinger, M. (2015), 'Smallholder experiences with dairy cattle crossbreeding in the tropics: from introduction to impact', *Animal: An International Journal of Animal Bioscience*, 9(1), pp. 150–7.

Schaeffer, L. R. (2006), 'Strategy for applying genome-wide selection in dairy cattle', *Journal of Animal Breeding and Genetics*, 123(4), pp. 218–23. Available at: http://www.ncbi.nlm.nih.gov/entrez/query.fcgi?cmd=Retrieve&db=PubMed&dopt=Citation&list_uids=16882088.

SNV, K. (2013), *Dairy Sector Policy Study and Capacity Needs Assessment of Stakeholder Associations*. Nairobi, Kenya.

South Africa (2014), *Summary of Livestock Diseases for South Africa Report for August 2014*. Available at: http://www.angoras.co.za/page/summary_of_livestock_disease_for_south_africa_report_for_august_2014.

© Burleigh Dodds Science Publishing Limited, 2017. All rights reserved.

Staal, J. S., Pratt, A. N. and Jabbar, M. (2008), *Dairy cattle Development for Resource Poor Part 2: Kenya and Ethiopia Dairy development Case Studies, Pro-poor Livestock Policy Initiative Working Paper N0. 44–2*. Report.

Staal, S. J. (2001), 'The competitiveness of smallholder dairy production: evidence from sub-Saharan Africa, Asia and Latin America', in D. Rangnekar and W. Thorpe (Eds), *South-South Workshop held at NDDB*. 13–16 March 2001, NDDB (National Dairy Development Board), Anand, India and ILRI (International Livestock Research Institute), pp. 250–64.

Syrstad, O. (1989), 'Dairy cattle cross-breeding in the tropics: performance of secondary cross-bred populations', *Livestock Production Science*, 23(1–2), pp. 97–106.

Tebug, S. F., Kasulo, V., Chikagwa-Malunga, S., Wiedemann, S., Roberts, D. J. and Chagunda, M. G. G. (2012), 'Smallholder dairy production in Northern Malawi: production practices and constraints', *Tropical Animal Health and Production*, 44(1), pp. 55–62.

Thornton, P., Herrero, M., Freeman, A., Mwai, O., Rege, E., Jones, P. and Mcdermott, J. (2007), 'Vulnerability, Climate change and Livestock – Research Opportunities and Challenges for Poverty Alleviation', *Open Access Journal published by ICRISAT*, 4(1), pp. 1–23. http://www.dfid.gov.uk/research/climate-change.asp.

Trivedi, K. R. (2002), 'Recommendations and summaries', in K. R. Trivedi (Ed.), *International Workshop on Animal Recording for Smallholders in Developing Countries, 20–23 October, 1997*. ICAR Technical Series 1, Anand, India.

Van Arendonk, J. A. M. (2011), 'The role of reproductive technologies in breeding schemes for livestock populations in developing countries', *Livestock Science*, 136, pp. 29–37.

Wudu, T., Kelay, B., Mekonnen, H. M. and Tesfu, K. (2008), 'Calf morbidity and mortality in smallholder dairy farms in Ada'a Liben district of Oromia, Ethiopia', *Tropical Animal Health and Production*, 40(5), pp. 369–76.

Zonabend, E., Okeyo, A. M., Ojango, J. M. K., Hoffmann, I., Moyo, S. and Philipsson, J. (2013), 'Infrastructure for sustainable use of animal genetic resources in Southern and Eastern Africa', *Animal Genetic Resources/Ressources génétiques animales/Recursos genéticos animales*, 53, pp. 79–93.

Zonabend, E., Okeyo, A. M., Ojango, J. M. K., Moyo, S. and Philipsson, J. (2010), 'Infrastructure for sustainable utilization of animal genetic resources', in *All Africa Conference on Animal Agriculture*. Addis Ababa, Ethiopia, pp. 1–11.

© Burleigh Dodds Science Publishing Limited, 2017. All rights reserved.

www.ingramcontent.com/pod-product-compliance
Lightning Source LLC
Chambersburg PA
CBHW050540270326
41926CB00015B/3319